Includes both Mac and PC software on the same CD-ROM

Published by

Astronomical Applications Department
U.S. Naval Observatory
3450 Massachusetts Avenue, N.W.
Washington, DC 20392-5420
United States of America

P.O. Box 35025
Richmond, VA 23235
Toll Free 1 (800) 825-7827
(804) 320-7016 • Fax (804) 272-5920
www.willbell.com

Published by Willmann-Bell, Inc., P.O. Box 35025, Richmond, Virginia 23235

Printed in the United States of America

Library of Congress Cataloging in Publication Data

Multiyear interactive computer almanac, 1800–2050 : 2.0 / Astronomical
Applications Department, U.S. Naval Observatory
p. cm.
Includes bibliographical references and index.
ISBN 0-943396-84-0
1. Astronomy--Data processing --Handbooks, manuals, etc. 2. MICA (Computer file) I. United States. Astonomical Applications Dept.

QB51.3.E43M84 2005
528--dc22 200550411106

05 06 07 08 08 10 9 8 7 6 5 4 3 2 1

Foreword

The U.S. Naval Observatory is one of the oldest scientific agencies in the country. It was established in 1830 as the Depot of Charts and Instruments. Its primary mission was to care for the U.S. Navy's chronometers, charts and other navigational equipment.

In 1844, as its mission evolved and expanded, the Depot was reestablished as the U.S. Naval Observatory and was located on the hill north of where the Lincoln Memorial now stands (i.e. Foggy Bottom). For nearly 50 years the Observatory remained at the Foggy Bottom location. During these years significant scientific studies were carried out such as speed of light measurements, the phenomena of solar eclipses and transit of Venus expeditions. The astronomical and nautical almanacs were started in 1855. In 1877, while working for the Naval Observatory, astronomer Asaph Hall discovered the two satellites of Mars.

However, by the 1890's, it was clear that the Naval Observatory had to move out of the city. Unhealthy conditions in the Foggy Bottom neighborhood had taken their toll. In 1893, after nearly 50 years at the site on the Potomac River, the U.S. Naval Observatory moved to its present location. At that time, this rural site was well outside the city in the countryside above Georgetown. The move not only provided better astronomical observing conditions, but also provided an opportunity to rethink old scientific programs and propose new ones. Along with the new programs such as daily monitoring of solar activity, the old functions of time keeping and telescopic observations were kept intact when the Observatory moved to the new site. The old Observatory in Foggy Bottom was declared a National Historic landmark in 1966.

Today, the U.S. Naval Observatory is one of the world's preeminent authorities in the areas of precise time and astrometry; determining and distributing the timing and astronomical data required for accurate navigation and fundamental astronomy.

The Astronomical Applications Department of the U.S. Naval Observatory computes, from fundamental astronomical reference data, the position, brightness, and other observable characteristics of celestial bodies, as well as the circumstances of astronomical phenomena. This information is of critical importance to navigation, military operations planning, scientific research, surveying, accident reconstruction, architecture, and everyday activities. The products of the AA Department—publications, software, algorithms, and expertise—are used by the U.S. Navy and the other armed services, civilian government agencies, the scientific research community, and the public. These products are regarded as benchmark standards throughout the world. The department also carries out a modest research program in celestial mechanics and positional astronomy to enable it to meet future needs.

Visit the U.S. Naval Observatory on-line at http://www.usno.navy.mil/.

Table of Contents

Acknowledgments

MICA is a product of the United States Naval Observatory. MICA 2.0 was developed by the following past and present members of the Observatory staff, listed in alphabetical order: John Bangert, William Harris, Paul Janiczek (retired), George Kaplan, Nancy Oliversen (project manager), Wendy Puatua, and William Tangren. The following individuals assisted with testing pre-release versions of MICA 2.0: Dan Benedict, Charles Booher, Geoff Chester, Steve Dick, Alan Downing, Ralph Gaume, Catherine Hohenkerk, Sethanne Howard, Paul Janiczek, Roger Mansfield, Jean Meeus, Robert Miller, Anthony Moller, Alice Monet, Roger Sinnott, Susan Stewart, Mark Stollberg, and Sean Urban. The following members of the USNO staff also contributed to the development of MICA 1.0: LeRoy Doggett (1941–1996), James Hilton, Jennifer Jeffries, Marie Lukac, and Steve Panossian. George Kaplan designed the MICA 2.0 start-up screen and icons.

The basic algorithms used in MICA for computing apparent, topocentric, and astrometric places of stars and planets are described by Kaplan, Hughes, Seidelmann, Smith, and Yallop (1989). MICA utilizes the 1980 International Astronomical Union (IAU) Theory of Nutation as described in Section 3.222 in the *Explanatory Supplement to The Astronomical Almanac*. Resolutions on precession, nutation, and Earth rotation adopted by the IAU in 2000 have not yet been implemented in MICA 2.0.

Historical values of "Delta-T" (= TT − UT1) were obtained from McCarthy and Babcock (1986). Modern values of Delta T were obtained from the IERS Rapid Service/Prediction Center web site. Delta T predictions to 2050 were supplied by Tom Johnson of USNO's Earth Orientation Department.

The physical ephemerides of the planets are based on values obtained from Seidelmann (2002). Expressions for the apparent visual magnitudes of the major planets (except Mercury and Venus) are from Harris (1961), while expressions for the magnitudes of Mercury and Venus are from Hilton (2003).

Ephemerides

MICA 2.0 utilizes the Jet Propulsion Laboratory (JPL) DE405 planetary and lunar ephemeris (Standish, 1998). Asteroid positions are based on the USNO/ AE98 ephemerides (Hilton, 1999). MICA 2.0 also utilizes the Lieske (1977a) Galilean satellite ephemerides, which were adapted from the original galsat50 FORTRAN routine (Lieske, 1977b).

Catalogs

Sean Urban (USNO) supplied the astrometric data for the MASC catalog from the USNO Washington Comprehensive Catalog Database (WCCD). The MASC astrometric data was derived from the Hipparcos and Tycho-2 catalogs which are products of the European Space Agency's Hipparcos mission. The MICA 2.0 Bright Star Catalog is the same catalog as given in section H of *The Astronomical Almanac*. Additional MASC and Bright Star catalog data were obtained from the SIMBAD astronomical database and from the catalog archive at the Centre de Données astronomiques de Strasbourg France (CDS). The MICA ICRF Radio Source Catalog is based on data published by Ma (1997) and has been updated to include the sources listed in the ICRF-Ext.2 Fey (2004).

Chapter 1

Introduction

The Multiyear Interactive Computer Almanac (MICA) 1800–2050 is a software system created especially for astronomers, surveyors, meteorologists, navigators, and others who regularly need accurate information on the positions and motions of celestial objects. It provides high-precision astronomical data for a variety of astronomical objects.

MICA computes many of the astronomical quantities that are tabulated in the *The Astronomical Almanac*—an annual publication prepared jointly by the Nautical Almanac Office of the U. S. Naval Observatory and the H. M. Nautical Almanac Office at the Rutherford Appleton Laboratory, United Kingdom.

Many of the quantities in *The Astronomical Almanac* are given with respect to the center of the earth (Geocentric). Astronomical quantities (like rise/set/twilight times) that depend on the observer's location are typically tabulated in *The Astronomical Almanac* at set intervals of latitude and/or date. The user often must perform additional calculations in order to obtain data for a specific location and/or time and date. MICA, unlike *The Astronomical Almanac*, computes this information for any specific date and location automatically. No additional computations should be necessary.

It is assumed that the user has some technical knowledge of positional astronomy and astronomical phenomena. MICA was designed for technical users who require high-precision positional data in tabular form. (Except for the new Sky Map feature, most MICA calculations produce only tabular data.) Other potential users may find programs that provide such astronomical data at lower precision or in graphical form to be more suitable for their needs.

1.1 MICA 2.0 Features

MICA can perform the following types of computations:

- Precise positions for the Sun, Moon, major planets, selected asteroids, selected bright stars, and cataloged objects (e.g., stars, quasars, galaxies, etc.) using external catalogs provided with the program. Ten different position types are available.
- Various astronomical time and reference system quantities (e.g., Sidereal Time, Nutation and Obliquity, Equation of the Equinoxes, Calendar/Julian Date conversions, and ΔT).

- Twilight, rise, set, and transit times for major solar system bodies, selected bright stars, selected asteroids and cataloged objects.
- Physical ephemerides useful for making observations of the Sun, Moon, and major planets. Both illumination and rotation parameters are available.
- Low-precision topocentric data describing the configuration of the Sun, Moon, major planets and selected asteroids at specified times and locations. MICA 2.0 also includes a Sky Map option as an aid in locating the objects.
- Solar and Lunar Eclipse and transits of Mercury and Venus visibility information.
- Positions of Jupiter and the Galilean Satellites, and offsets of the Satellites from Jupiter.
- Dates and circumstances of various astronomical phenomena (Solstices and Equinoxes, Moon phases, conjunctions, oppositions, greatest elongations of Mercury and Venus). A phenomena search feature is also available which generates a table similar to the 'Diary of Phenomena' tables contained in section A of *The Astronomical Almanac*.

1.2 History and Enhancements

The first version of MICA (1.0) was published in 1993 by the National Technical Information Service (NTIS). In 1995, Willmann-Bell, Inc. published MICA 1.5. Two versions of MICA were available on the same distribution CDROM. The PC version was an MS-DOS based application for IBM-PC compatible computers and could be used on Windows systems (Windows 3.x, Windows 95, Windows 98), but the interface did not utilize modern Windows capabilities. The Macintosh version was designed for Macintosh Plus systems running System 6.0.2 or higher.

Previous users of MICA will find many additions and enhancements to MICA 2.0. All of the original MICA 1.0 computations can also be completed with MICA 2.0. The following changes and/or modifications have been added to MICA 2.0:

- Computations can be performed for the time period from 1800 to 2050 thus enabling historical calculations as well as predictions several years into the future. (MICA 1.5 covered the time period from 1990 to 2005.)
- The solar, lunar and planetary ephemeris is based on the JPL DE405 ephemeris (MICA 1.5 utilized JPL DE200).
- MICA 2.0 can now do various computations (positions, rise/set, configurations, conjunctions, oppositions) for 15 of the largest asteroids. These computations utilize the USNO/AE98 ephemerides.
- A Sky Map option has been included with the configuration section, which displays topocentric positions of the Sun, Moon, major planets, se-

lected asteroids, and bright stars.

- A selection of astrometric catalogs have been updated and included with the MICA 2.0 software. Position and rise/set calculations can be performed utilizing the catalog objects.
- The MICA rise/set computation code has been revised. Double transit information is now included, and the computations are more reliable at latitudes above 67° North latitude and below 67° South latitude.
- MICA 2.0 can now compute the dates of various astronomical phenomena (Solstices/Equinoxes, Moon phases, conjunctions, oppositions, and greatest elongations).

1.3 Computer System Information

MICA 2.0 has been designed for modern computers running the PC/Windows® or Mac OS operating systems.

1.3.1 Minimum System Requirements — PC version

- PC-compatible 200 MHz Pentium or higher
- Windows®, 98, Millennium Edition, NT® 4.0 (with Service Pack 4 or later), 2000, or XP operating system
- Internet Explorer version 4.0 (5.0 recommended)
- 64 MB of RAM (128 MB recommended)
- 135 MB of hard disk space (848 KB on the C: drive)
- CD-ROM drive for installation
- VGA or higher-resolution monitor (with screen area set to at least 640 x 480 pixels and 256 colors)

1.3.2 Minimum System Requirements — Macintosh version

- Any PowerPC-based Mac running Mac OS 9.2.2 or higher.
- For best performance, a G3 or faster processor running Mac OS X is recommended.
- Systems running Mac OS 9.2.2 require CarbonLib 1.6 installed. CarbonLib 1.6 can be downloaded from the Apple Web site (www.apple.com/support/downloads). OS X systems do not need to install Carbonlib.
- 200 MB of hard disk space.

1.4 Installation and Startup

The MICA 2.0 CD-ROM is a hybrid CD containing the both the Windows and Macintosh versions of the product.

1.4.1 Windows Automatic Installation

Insert the MICA CD-ROM into the CD-ROM drive. The MICA installation program should begin automatically. [If the MICA installation program does not start within 30 seconds, follow the instructions below for Windows manual installation.] Follow the prompts through the installation wizard and restart the machine, if required. MICA 2.0 is now ready to use.

1.4.2 Windows Manual Installation

Double-click on the 'My Computer' icon, then double-click on the CD-ROM drive (usually D:) icon to view the contents of the MICA CD-ROM. Next, double-click on 'setup.exe' to start the installation program.

1.4.3 Windows Startup

Either double-click on the MICA icon on the desktop or select the MICA program from the Start menu.

1.5 Macintosh Installation and Startup

Drag the enclosed 'MICA 2' folder to the Applications folder, or other desired location, on your harddrive. Mac OS X users may want to drag the MICA 2 application icon to the dock for easy access to the application. To start the application, double click on the MICA application icon, or click once on the icon in the dock.

1.5.1 Macintosh Help

To access the Macintosh Help, simply select the Help menu. For most Mac operating systems (OS 9.2.2 and OS X), the Help utility should start up automatically. However, for some selected OS versions, the help window may be blank. If the help window is blank, then go to the MICA 2 applications folder and double click on the 'MICA.help' file. This should bring up the MICA help viewer.

1.6 Disclaimer

1.6.1 Disclaimer of Liability

The MICA program, documentation, and associated data tables are provided to the user AS IS without warranty of any kind, either expressed or implied, including but not limited to the implied warranties of merchantability and fitness for a particular purpose. The entire risk as to the results, quality, and performance of the program, documentation, and associated data tables is with the user. Neither Willmann-Bell, Inc., the Department of the Navy, the U. S. Naval Observatory, its employees, nor the authors of the program warrant or guarantee the program, documentation, and associated data tables in terms of correctness, accuracy, reliability, currentness, or otherwise.

Neither Willmann-Bell, Inc. or the Department of the Navy, the U. S. Naval

Observatory, its employees nor anyone else involved in the creation, production, or official distribution of this program, its documentation, and associated data tables shall be liable for any direct, indirect, consequential, or incidental damages arising out of the use, the results of use, or inability to use this product even if the U. S. Naval Observatory or Willmann-Bell, Inc., have been advised of the possibility of such damages or claim. Should this program, documentation, or associated data tables prove defective, the user assumes the entire cost of all necessary servicing, repair, or correction.

Furthermore, Willmann-Bell, Inc. and the U. S. Naval Observatory are under no obligation to modify, support, or enhance this product. MICA, its documentation, and its associated data tables may be withdrawn from availability by the U. S. Naval Observatory or Willmann-Bell, Inc. at any time.

1.6.2 Disclaimer of Endorsement

The mention of any product, corporation, manufacturer, or service in MICA or its documentation does not constitute an endorsement by the Department of the Navy or the U. S. Government.

1.7 Technical Support

Have a question about MICA? Before contacting us, please check the MICA help system described above, and check the MICA Web site at the address given below. If you still need help, contact us by e-mail at help@aa.usno.navy.mil. Comments and suggestions regarding MICA are also welcome. Send all written correspondence to:

U.S. Naval Observatory
Code: AA/MICA
3450 Massachusetts Avenue, NW
Washington, DC 20392-5420

We regret that we cannot reply to telephone inquiries

1.8 More Information

For additional MICA 2.0 information, announcements, and minor updates, visit: http://aa.usno.navy.mil/software/mica/micainfo.html.

Visit the Astronomical Applications Department Web site, http://aa.usno.navy.mil/, for information on our other products, and for a wide array of astronomical information and data services.

Chapter 2

Astronomical Background

This chapter provides a brief introduction to some of the astronomical concepts, and techniques used throughout MICA. See also the Glossary for a list of some of the terminology used in MICA. Further technical information can be obtained from *The Explanatory Supplement to The Astronomical Almanac*[1] and the appendices to *The Astronomical Almanac*.

2.1 Position Calculations

There seems to be a bewildering number of ways of calculating the positions of celestial bodies. This variety has built up over the centuries to meet specific needs. The need for more than one system arises because the platform from which we observe is in constant motion. For centuries observations have been recorded in coordinates that seemed natural to an Earth-based observer. However, the laws of mechanics are more easily expressed in an inertial coordinate system with an origin at the center of mass of the bodies of interest. As our theoretical understanding and the accuracy of our observations have increased, many approaches have been developed to deal with this problem.

When positions of celestial objects are needed, several parameters specify the calculations and the type of resulting Position. On the PC, these parameters are Place, Origin, and Frame. On the Mac, five parameters (Origin, Place, Reference Epoch, Reference Plane, Coordinates) are used to specify the position calculations. The MICA position parameters are explained below.

2.1.1 Place

The *Place* parameter is used to denote standard forms of astronomical calculations. Essentially, it specifies the complexity of the model of the observing geometry; i.e., how much light-propagation physics is used. The choices are:

- **Geometric** A geometric place is formed simply by a vector difference of the instantaneous positions of the object and the observer, as obtained from catalog data or the planetary ephemeris. No light-propagation physics is applied. Geometric places are basic data useful for investigating motions in the solar system. The observer is usually a hypothetical one

[1] Seidelmann, P.K., 1992, University Science Books

located at the center of the Sun, Earth, or the solar system barycenter.

- **Astrometric** An astrometric place is a geometric place which, for solar system objects, has been corrected for light propagation time. (Light-time computations are never done for stars; it is assumed that the catalog positions and proper motions of stars implicitly include light-time and its derivative.) It is comparable to the positions of stars that are published in catalogs and is therefore useful in plotting the positions of solar system objects on star charts.
- **Apparent** A calculated apparent place corresponds most closely to the observed position of an object on the celestial sphere. The aberration of light (due to the velocity of the observer) and the relativistic bending of light (due to the Sun's gravitational field) are taken into account. For solar system objects, light propagation time is also included.

2.1.2 Origin

The *Origin* parameter specifies the origin of the coordinate system which is the position of the real or hypothetical observer. The choices are:

- **Topocentric** positions as they are observed from the surface of the Earth are designated as topocentric. They depend on the geographic location of the observer, but do not include atmospheric refraction or polar motion. Topocentric positions come closest to specifying what an observer sees at a specific location on the surface of the Earth.
- **Geocentric** coordinates specify the position of a celestial object as it would be observed from the center of the Earth. They are used in published almanacs, since for many applications they differ little from topocentric coordinates.
- **Heliocentric** coordinates pertain to a hypothetical observer at the center of the Sun. Such coordinates are useful for describing motion in a simplistic model of the solar system with the Sun at the center.
- **Barycentric** The origin of barycentric coordinates is located at the center of mass of the solar system, which is close to and sometimes inside the Sun. Although the barycenter is in motion around the center of the galaxy, for calculating solar system ephemerides it is considered to define the origin of an inertial reference frame.

2.1.3 Reference Epoch

Two conventions are used by MICA, J2000.0 and date.

- **J2000** The standard epoch for star catalogs is denoted J2000.0. When this epoch is used, the reference plane and zero point are oriented to their mean position at 2000 January 1, 12h (JD 245145.0).
- **Date** The epoch of "date" implies that the reference frame is oriented to the true equinox and reference plane of the date for which the data is cal-

culated. Coordinates on a reference frame of date are relevant to the actual observations, since the true equator of date is an extension of the Earth's equator.

2.1.4 Frame

The *Frame* specifies the orientation of the coordinate axes of the reference frame. The Frame can be expressed as two components, Reference Epoch (date or J2000.0), and Reference Plane (equator, ecliptic or horizon). The following combinations are available in MICA 2:

- **Equator of Date** (RA and Dec) This coordinate system is oriented with its *xy*-plane parallel to the true instantaneous Earth equator at the time of observation, and its *z*-axis pointing toward the true instantaneous north celestial pole. The *x*-axis points toward the true instantaneous equinox. This coordinate system is useful for expressing the positions of observed objects with respect to Earth-based equatorially-mounted instruments.
- **Ecliptic of Date** This coordinate system is oriented with its *xy*-plane parallel to the ecliptic (Earth's orbital plane) at the time of observation, and its *z*-axis pointing toward the north ecliptic pole. The *x*-axis points toward the true instantaneous equinox. This coordinate system is useful for expressing the relative positions of several solar system objects, as seen from Earth.
- **Equator of J2000** This coordinate system is oriented with its *xy*-plane parallel to the mean Earth equator at epoch J2000.0, and its *z*-axis pointing toward the mean north celestial pole of J2000.0. The *x*-axis points toward the mean equinox of J2000.0. This coordinate system is used for expressing the positions of stars in catalogs and planets in basic solar system ephemerides.
- **Ecliptic of J2000** This coordinate system is oriented with its *xy*-plane parallel to the ecliptic (Earth orbital plane) at epoch J2000.0, and its *z*-axis pointing toward the north ecliptic pole. The *x*-axis points toward the mean equinox of J2000.0. This coordinate system is useful for expressing the relative positions of solar system objects as given in basic solar system ephemerides.
- **Local Horizon** This coordinate system is a left-handed system oriented with the its *xy*-plane parallel to the horizon of an Earth-based observer, and its *z*-axis pointing toward the zenith. (Assuming geodetic longitude and latitude have been entered, the zenith is the normal to the Earth's reference ellipsoid at the observer's location.) The *x*-axis points toward true north and the *y*-axis points toward true east.
- **Equator of Date** (LHA and Dec). This coordinate system is oriented with its *xy*-plane parallel to the true instantaneous Earth equator at the time of observation, and its *z*-axis pointing toward the true instantaneous north celestial pole. The *x*-axis points toward the point where the local

Table 2.1 Common Position Types	
Astronomical Activity	**Position Types (Place, Origin, Frame; Reference Epoch, Reference Plane)**
Pointing an altitude/azimuth telescope; Calibrating surveying instruments; or Reducing a sextant observation.	Apparent, Topocentric, Horizon; Date, Spherical
Pointing an equatorial telescope.	Apparent, Topocentric, Equator of Date; Date, Equator Apparent, Geocentric, Equator of Date; Date, Equator
Plotting planetary positions on a star chart.	Astrometric, Geocentric, Equator of J2000.0; J2000.0, Equator
Analyzing planetary motions.	Geometric, Heliocentric, Equator of J2000.0; J2000.0, Ecliptic Geometric, Barycentric, Equator of J2000.0; J2000.0, Equator

meridian and the celestial equator intersect. The local hour angle (LHA) is useful for pointing telescopes having an hour angle setting circle. Note that on the Mac this option is obtained by selecting *Hour Angle* under the *Coordinates* parameter.

2.1.5 Coordinates

There are three choices on how the coordinates are to be tabulated. On the Mac this is specified by the Coordinates parameter. On the PC the coordinates are selected automatically based on the previous parameter choices (*Place*, *Origin*, and *Frame*).

- **Spherical** Positions are given in terms of angles on the celestial sphere.
- **Rectangular** Positions of an object are with respect to three mutually perpendicular axes (x, y, and z) that cross at some specified origin, in units of Astronomical Units.
- **Hour Angle** Positions are given in Local Hour Angle (see Glossary) and Declination

The ensemble of position parameter choices (Place, Origin, and Frame) describe 60 possible position types. However, some combinations of position parameters make little sense, and most are simply of no practical use. MICA deactivates certain position combinations and limits the choices to 10 position types which are in reasonably common use. The type of position calculation possible varies with the object (e.g., it makes little sense to do a heliocentric position for the Sun!).

Table 2.1 on page 10 lists some common astronomical activities and the corresponding MICA position parameter combinations.

2.2 The Measurement of Time

2.2.1 Fundamental Time Scales

Astronomical observations made from the surface of the Earth are tied to the state of the Earth's rotation, which forms the basis for one kind of measure of time. However, in the 20th century it was discovered that the Earth does not rotate at a constant rate. More precise measures of time can be made by observing atomic or dynamical processes. There is now a family of time scales, with precise technical definitions, that replaces what was once called simply Universal Time. In MICA we explicitly use two time scales. TT is an atomic time scale corresponding to an idealized clock on the surface of the Earth, and UT1 is based on the Earth's rotation.

TT (Terrestrial Time) is the time scale that would be kept by an ideal clock on the geoid — approximately, sea level on the surface of the Earth. Since its unit of time is the SI (atomic) second, TT is independent of the variable rotation of the Earth. In practice TT is derived from International Atomic Time (TAI), a time scale kept by real clocks on the Earth's surface, by the relation TT = TAI + $32\overset{s}{.}184$. It is the time scale now used for the precise calculation of future astronomical events observable from Earth. TT is considered to be a continuation of the old Ephemeris Time (ET) scale. Note that, TT time in MICA 1.x was labeled as TDT for Terrestrial Dynamical Time. The International Astronomical Union recommended in 1991 that TDT be renamed simply TT, which removes the word "dynamical" from the name.

UT1 (Universal Time 1) is a time scale that is based on the rotation of the Earth. UT1 is related to TT by the formula UT1 = TT – ΔT, where ΔT is determined from astronomical observations (its value is now about one minute). In current practice, UT1 is defined by its relationship to sidereal time, an observable quantity. Since we cannot accurately predict the future behavior of the Earth's rotation, there is always some uncertainty in extrapolating UT1 into the future. In MICA 2.0 ΔT is obtained by interpolating a series of values (actual determinations for past dates; plus predictions for future dates) produced by the U. S. Naval Observatory's Earth Orientation Department. Because predictions of ΔT are necessarily imprecise, UT1 cannot be used for high precision calculations for future dates. For example, UT1 is not suitable for the calculation of dynamical phenomena such as planetary motions.

TDB (Barycentric Dynamical Time) is the time scale that would be kept by an ideal clock, free of gravitational fields, co-moving with the solar system barycenter. It is always within 2 milliseconds of TT, the difference caused by relativistic effects. Although MICA uses TDB internally, it is not used as the time basis for any MICA input or output parameters. TDB is the time scale now used for investigations of the dynamics of solar system bodies.

UTC (Coordinated Universal Time) time scale is the basis of international civil time keeping. For most places, civil time is simply an integral number of hours offset from UTC. UTC is a hybrid time scale: its rate is the same as that of

International Atomic Time (TAI), but its epoch is occasionally adjusted in one-second steps (leap seconds) to keep it within 0.9 second of UT1. Since it is impossible to predict when these step adjustments will be introduced, MICA does not use UTC. UTC time is disseminated by GPS, Loran-C and other broadcast systems. Note that GPS system time is referenced to the Master Clock (MC) at the USNO and steered to UTC(USNO) from which system time will not deviate by more than one microsecond.

Values of UT1 – UTC are published weekly on the IERS Rapid Service/Prediction Center web site[1] (in the U. S. by the Earth Orientation Department of the U. S. Naval Observatory[2]). Transmitted time signals of UTC include a coded value of ΔUT1, the approximate difference between UTC and UT1. UTC can therefore be converted to UT1 (the latter used in MICA) through the relation UT1 = UTC + ΔUT1.

2.2.2 Ephemeris Meridian

The ephemeris meridian is a concept closely related to the time scale TT which is independent of the (irregular) rotation of the Earth. The ephemeris meridian can be thought of as occupying the position that the Greenwich meridian would occupy if the Earth rotated uniformly according to standard formulas. The ephemeris meridian is 1.002738 ΔT east of (ahead of) the Greenwich meridian (currently less than 30 km at the equator). The ephemeris meridian defines a system of ephemeris longitudes. In MICA, choosing TT as the time scale automatically shifts the geographic longitude system to that defined by the ephemeris meridian. In this way it is possible to compute topocentric phenomena in a self-consistent system which is independent of the Earth's variable rotation.

2.2.3 Delta T (ΔT)

ΔT is the difference between Terrestrial Time (TT) and Universal Time (UT): ΔT = TT – UT1. MICA 2.0 contains an internal table of historical and predicted ΔT values and dates. MICA 2.0 determines the ΔT value for a specific date and time by linear interpolation between adjacent entries in this table. The MICA ΔT table data is derived from three sources. For the years from 1800 through 1972, semi-annual ΔT values were obtained from McCarthy and Babcock (1986). For the years from 1973 to late 2004, monthly ΔT values were derived from information contained on the IERS Rapid Service/Prediction Center web site, which is maintained by the USNO Earth Orientation Department. Future (2004–2050) values of ΔT are based on predictions by Johnson (2004). The estimated error of these predictions increase from ±0.004s in 2005 to ±24s in 2050. Note that the MICA 2.0 ΔT file will be periodically updated and made available via the USNO MICA web site.

---

[1] http://www.iers.org/

[2] http://maia.usno.navy.mil/

2.2.4 Julian Date

The Julian date provides a convenient, continuous count of dates through recorded history. It is the interval of time in days and fraction of a day since Greenwich noon on 1 January 4713 BCE, Julian calendar. The starting date of MICA, 1 January 1800, 0h, is JD 2378496.5. Sometimes it is necessary to state the fundamental time scale on which the Julian date is based; e.g., the standard epoch J2000.0, 1 January 2000, 12h, is JD 2451545.0 (TDB).

2.2.5 Sidereal Time

Sidereal time is a direct, observable measure of the rotation of the Earth with respect to the stars. The sidereal time at any location is equal to the apparent right ascensions of stars transiting the local meridian. Standard formulas relate sidereal time to UT1.

2.3 Solar System Ephemerides

MICA 2.0 utilizes the Jet Propulsion Laboratory (JPL) DE405 ephemerides for position calculations of the Sun, Moon and major planets. The DE405 ephemerides provide barycentric equatorial rectangular coordinates for the period 1600 to 2201 (Standish, 1998). MICA 2.0 includes a subset of the DE405 period, from 1800 to 2050. Note that the MICA 1.x ephemeris was derived from DE200 and covered the period from 1990 to 2005. The MICA 2.0 extended ephemeris enables historical calculations as well as predictions several years into the future.

The reference frame for the basic DE405 ephemerides is the ICRF; the alignment onto this frame has an estimated accuracy of 1–2 milliarcseconds. The ephemerides have been developed in a barycentric reference system using a barycentric coordinate time scale T_{eph}, which in MICA is assumed to be equivalent to TDB.

2.4 Asteroid Ephemerides

MICA 2.0 utilizes the USNO/AE98 ephemerides for various asteroid position computations (Hilton, 1999). These computations are included in the MICA 2.0: Positions, Rise/Set/Transit, Configurations (Asteroids and Sky Map), and Phenomena (Conjunctions, Oppositions) tasks. MICA 2.0 performs computations for fifteen of the largest asteroids: Ceres, Cybele, Davida, Eunomia, Europa, Flora, Hebe, Hygiea, Interamnia, Iris, Juno, Metis, Pallas, Psyche, and Vesta.

2.5 MICA External Catalogs

MICA has the ability to do Position and Rise/Set/Transit time calculations for stars or objects listed in external catalog files. The following external catalogs have been included with the MICA 2.0 CDROM.

2.5.1 MICA Astrometric Source Catalog (MASC)

The MASC catalog contains all stars recorded in the Hipparcos and/or the Tycho-2 catalogs with a Tycho magnitude (V_T) brighter than 9.5 mag. There are 232317 entries in this file. Most entries are listed in the Tycho-2 Catalog and approximately half are also listed in the Hipparcos Catalog. The Hipparcos catalog has a median astrometric precision of about 1 milliarcsec, while the Tycho catalog has a median astrometric precision of about 7 miliarcsec for objects with V< 9 mag. The positions and proper motions are given within the ICRS, for the mean equinox and equator of J2000. The MASC catalog was compiled from data contained in the USNO Washington Comprehensive Catalog Database. Additional information (especially alternate object name information) was also obtained from the SIMBAD object database and from the catalog archive at the Centre de Données astronomiques de Strasbourg France (see the CDS web site). The Hipparcos and Tycho-2 catalogs are products of the European Space Agency's Hipparcos mission, which operated from November 1989 to March 1993. A Vmagnitude flag, Spectral Type, Multiple flag, and Variable flag are also displayed in the MASC catalog output.

2.5.2 ICRF Radio Source Catalog

The MICA 2.0 ICRF Radio Source Catalog contains all the 608 extragalactic radio sources listed on the ICRF web site plus the additional 109 new radio sources listed in the ICRF-Ext.2 Fey (2004). The objects in the ICRF are classified in four categories: defining, candidate, "other," and new sources, all four of which are included in the MICA 2.0 catalog. The catalog data are derived from Ma (1997), and the candidate, "other," and new source positions have been updated to Fey (2004). Positions are referred to the mean equator and equinox of J2000. Additional object name information, apparent visual magnitude (V), object category (C), object type (QSO, Galaxy, BL LAC, other), redshift (Z), and source flux density (Janskys at 5 GHz) were also obtained from the ICRF web site and are included in the MICA catalog.

2.5.3 Navigational Star Catalog

The MICA Navigational Star List contains the 59 bright stars commonly used for Celestial Navigation. The stars are the same ones as listed in The Nautical Almanac with the addition of Polaris and Sigma Octantis. The MICA 2.0 catalog contains the object name and the object number as listed in The Nautical Almanac. The positions and proper motions are given within the ICRS, for the mean equinox and equator of J2000.

2.5.4 AsA Bright Star Catalog

This catalog contains 1467 stars which are the same as those listed in the Bright Stars catalog in Section H of *The Astronomical Almanac*. The mean positions for all stars in this catalog were obtained from the Hipparcos Catalogue and are referred to the equator, equinox, and epoch of J2000.0. The MICA 2.0 Bright Star catalogue contains all the fields as are listed in the AsA Bright Star catalog. The common or historical object names have also been included in the MICA version of the Bright Star catalog and were obtained from the object database at the SIMBAD web site. For further information see the Bright Star Catalog field description and/or Section L of *The Astronomical Almanac*. Stars were included in the catalogue according to the following criteria:

1. all stars of visual magnitude 4.5 or brighter, as listed in the fifth revised edition of the Yale Bright Star Catalog (BSC);
2. all FK5 stars brighter than 5.5;
3. all Morgan-Keenan (MK) atlas standards in the BSC Morgan, et. al., (1978) and Keenan and McNeil (1976);
4. all stars matching criteria 1–3 and also listed in the Hipparcos Catalogue.

2.5.5 Messier Object Catalog

This catalog is unchanged from MICA 1.x. It lists 108 deep-sky objects by increasing Messier number, and the positions of all objects are referred to the equator, equinox, and epoch of J2000.0.

2.5.6 External Catalogs

This selection is available to support a previously imported MICA 1.x catalog, including a user-created catalog. For details on importing a MICA 1.x catalog see the File menu.

2.6 MICA 2.0 Bright Stars List

MICA contains an internal list of 22 stars brighter than about mv = 2.0 which are used in the Positions and Rise/Set/Transit calculations. These stars are also in the MASC and AsA Bright Star catalogs, as well. The following stars are included in this list:

Achernar	Hadar
Adhara	Mimosa
Aldebaran	Polaris
Altair	Pollux
Antares	Procyon
Arcturus	Regulus
Betelgeuse	Rigel
Canopus	Rigil Kentaurus
Capella	Sirius
Deneb	Spica
Fomalhaut	Vega

Chapter 3

Menus

The purpose of this and the next two chapters is to familiarize the user with MICA's interface. Overall, the arrangement of the menus for both the PC and Mac follow the same general layout. Where there are differences it is almost always to accommodate the conventions of the particular operating system. Here we will use the PC as our general frame of reference and note those instances where they differ from the Mac. Chapter 4 will be devoted to Windows dialog boxes and Chapter 5 to the Mac.

3.1 File Menu Commands

3.1.1 For the PC

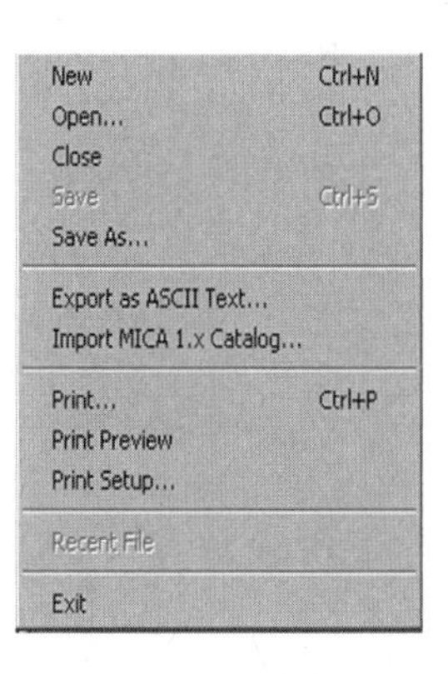

New Creates a new MICA worksheet document.

Open Opens an existing document (either MICA worksheet or Sky Map file).

Close Closes an opened document.

Save Saves an opened document. If that document has never been saved before, MICA will prompt to save it using the current default name, in the current default location. One has the option to change the name and /or location, if desired. If the document had previously been saved, the Save option will save it again, and the user will NOT be prompted to change the name or save location.

Save As Saves an opened document. For a previously unsaved document, the action of the Save As function is identical to the Save function. If the document HAD been saved before, the Save As function will behave as it does for an unsaved document.

Export as ASCII text Saves the current MICA worksheet data to an ASCII text document.

Import MICA 1.x Catalog Converts a MICA 1.x catalog into the new format for use with MICA 2.0 Browse to the location of the MICA 1.x catalog (the file extension must be .CAT), select the file, and click the 'Open' button. A message will confirm the success of the import and put the new catalog in the MICA application folder alongside the other MICA 2.0 catalogs.

Print Prints a document:

Print Preview Displays the document on the screen as it would appear printed.

Print Setup Displays the system dialog box associated with your default printer.

Exit Exits MICA.

3.1.2 For the Mac

New Creates a new MICA document window.

Open Opens an existing text document.

Close Closes an opened document.

Save Saves an opened document using the same file name. If the document has not been previously saved, then the user will be prompted to provide a file name.

Save As Saves an opened document to a specified file name.

Revert Loads the last saved version of the current document.

Import Catalog Converts a MICA 1.x catalog into the new format for use with MICA 2.0. Browse to the location of the MICA 1.x catalog, select the file, and click the 'Choose' button. A message will confirm the success of the import. The new catalog is placed inside the "Catalogs" folder alongside the other MICA 2.0 catalogs.

Page Setup Displays the system dialog box associated with the selected printer.

Print Prints a document to the selected printer

3.2 Edit Menu Commands

3.2.1 For the PC

Undo/Redo For Windows 2000 and XP, Undo removes the last text typed or pasted into the MICA worksheet. Redo reverses this. Undo/Redo DO NOT WORK on text put into the MICA worksheet *by the program itself.* In Windows 98, Undo removes everything from the worksheet, Redo puts it all back. If a user cuts from the worksheet, then Undo/Redo only works on what was cut or pasted from the clip board. This is true for all versions of Windows.

Cut Deletes data from the MICA worksheet and moves it to the clipboard.

Copy Copies data from the MICA worksheet to the clipboard.

Paste Pastes data from the clipboard into the MICA worksheet.

Select All Selects all the text in the MICA worksheet.

Find Searches for specified text in the MICA worksheet.

Find Next Finds and selects the next occurrence of the text specified in the Find what box.

Replace Searches for and replaces specified text in the MICA Worksheet.

3.2.2 For the Mac

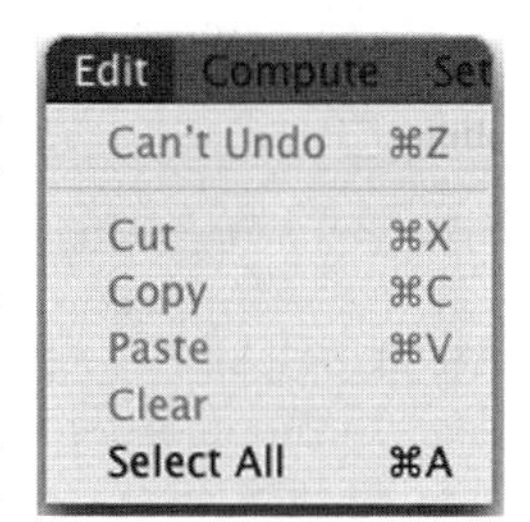

Undo Reverses the previous change to the MICA document window. Note that after a calculation has been performed 'Undo Typing' will just remove the last line of the document window.

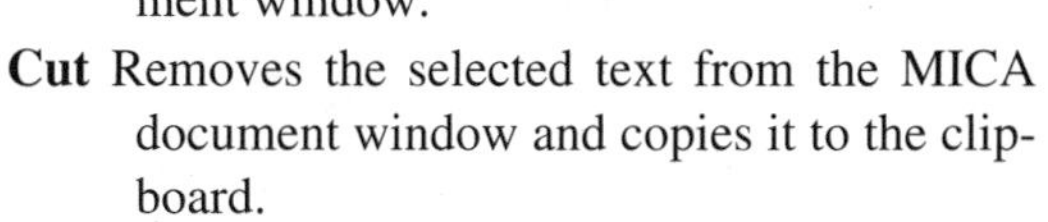

Cut Removes the selected text from the MICA document window and copies it to the clipboard.

Copy Copies the selected text from the MICA document window to the clipboard.

Paste Pastes data from the clipboard into the MICA document window.

Clear Removes the selected text from the MICA document window.

Select All Selects all the text in the MICA document window.

3.3 View Menu Commands (PC)

Tool Bar Shows or hides the tool bar.

Status Bar Shows or hides the status bar.

3.4 Calculate/Compute Menu

The MICA Calculate (PC) or Compute (Mac) menu offers the following commands:

3.4.1 Positions

The Positions section of MICA generates positions for the Sun, Moon, major planets, selected asteroids, and selected bright stars. It also has the capability to do position calculations for stellar or star-like objects listed in external catalogs.Ten different position types are available (see Table 3.1).

To obtain a tabular listing of positions on the PC, select Positions from the MICA Calculate menu. Next select the desired Object and Position Type in the

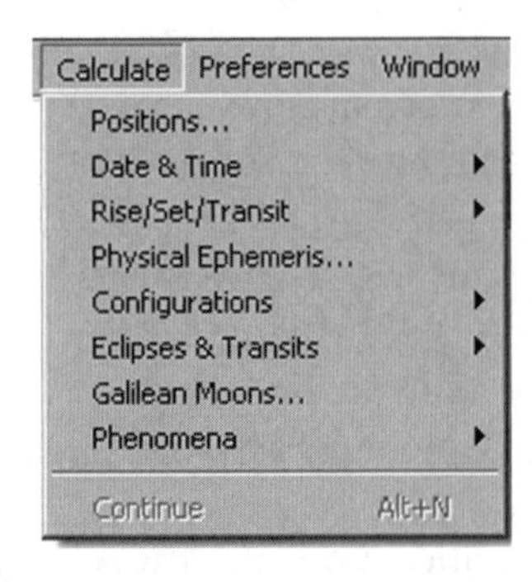

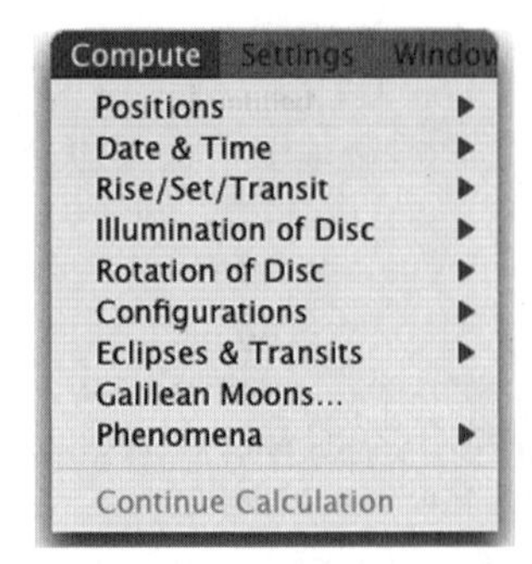

Position dialog box. If an asteroid, bright star, or a catalog position is wanted, the desired object(s) should be specified in the respective asteroid, bright star, or catalog dialog boxes. Next, enter the date and time, number of repetitions of the calculation, the Time System, and observer's location (for topocentric calculations only) in the Date/Time/Location dialog box.

To obtain a tabular listing of positions on the Mac, select Positions from the MICA Compute menu and then select the desired object from the displayed list. Next a position dialog box will be displayed, with separate views for entry of the Coordinate System information, Date/Time, Repetitions, Location (for topocentric calculations only), and Output format. If an asteroid or a bright star position is selected, then an additional view is present for selection of the specific object. If a catalog position is selected, then two extra views (Catalog and Search) are present, one for selection of the catalog and the second to specify the catalog search criteria.

3.4.1.1 MICA Position Types

There are up to ten different position types available in MICA 2.0. The exact type of position calculation possible varies with the object. See Table 3.1 for a summary of the MICA position types.

3.4.1.2 Position Output Tables

The heading for each table produced using the Positions command contains the object name (except for catalog objects where the name is in the table body), the type of place (Apparent, Geometric and/or Astrometric), the origin (Barycentric, Geocentric and/or Topocentric), and the Frame/Date information (True Equator and Equinox of Date, Mean Equator and Equinox of J2000, Mean Ecliptic and Equinox of J2000.0, Local Zenith and True North, etc.). If a topocentric origin was selected, the observer's location (latitude, longitude, and height above sea level) and longitude reference (either the Greenwich meridian for the UT1 timescale or Ephemeris meridian for the TT time scale) will appear in the heading as well. The time scale (UT1 or TT) and date format (Julian or calendar) are given in the headings. Rectangular coordinates and distances are tabulated in astronomical units (AU), except for the geocentric positions of the Moon, where kilometers (km) are used.

Table 3.1 MICA Position Types

Position Type (Place, Origin, Frame)	Objects	Tabulated Coordinates
Geometric Barycentric Equator of J2000	Sun Moon Earth Asteroids Planets	X, Y, and Z
Geometric Heliocentric Equator of J2000	Moon Earth Asteroids Planets	X, Y, and Z
Geometric Heliocentric Ecliptic of J2000	Moon Earth Asteroids Planets	Ecliptic longitude, Ecliptic latitude, and distance
Geometric Geocentric Equator of J2000	Sun Moon	X, Y, and Z
Astrometric Geocentric Equator of J2000	Sun Moon Asteroids Planets (except Earth) Stars Catalog objects	Right ascension, Declination Distance (for solar system objects).
Apparent Geocentric Equator of Date	Sun Moon Asteroids Planets (except Earth) Stars Catalog objects	Right ascension, Declination, Distance (for solar system objects) Equation of Time (Sun and UTI time only)
Apparent Geocentric Ecliptic of Date	Sun Moon Asteroids Planets (except Earth) Stars Catalog objects	Ecliptic Longitude, Latitude, Distance (for solar system objects)
Apparent Topocentric Equator of Date (RA, Dec)	Sun Moon Asteroids Planets (except Earth) Stars Catalog objects	Right Ascension Declination Distance (for solar system objects)
Apparent Topocentric Local Horizon	Sun Moon Asteroids Planets (except Earth) Stars Catalog objects	Zenith angle Azimuth angle Distance (for solar system objects)
Apparent Topocentric Equator of Date (LHA, Dec)	Sun Moon Asteroids Planets (except Earth) Stars Catalog objects	Local Hour Angle Declination

For catalog objects, several additional catalog-dependent fields are included in the position table output. Several sample position tables are shown below. Note that the right sides of the catalog tables have been truncated for printing purposes.:

Apparent Geocentric Positions
True Equator and Equinox of Date

MICA Astrometric Stellar Catalog

Object ID	Date	Time	Right Ascension	Declination	V	V	Spectral Type	Multiple	Variable
	(UT1)					Flag		Flag	Flag
		h m s	h m s	° ′ ″	mag				
REGULUS	2004 Sep 01	00:00:00.0	10 08 35.458	+ 11 56 53.18	1.36	H	B7V	1	

Apparent Geocentric Positions
True Equator and Equinox of Date

ICRF Radio Source Catalog

Object ID	Date	Time	Right Ascension	Declination	V	Cat	Object Type	Z	Flux (6cm)
	(UT1)								
		h m s	h m s	° ′ ″	mag				Jy
0146+056	2004 Sep 01	00:00:00.0	1 49 37.242	+ 5 57 23.28	20.0	C	Q	2.345	1.17

Apparent Geocentric Positions
True Equator and Equinox of Date

Navigational Star Catalog

Object ID	Date	Time	Right Ascension	Declination	V
	(UT1)				
		h m s	h m s	° ′ ″	mag
Hamal	2004 Sep 01	00:00:00.0	2 07 26.270	+ 23 29 05.70	2.0

Apparent Topocentric Positions
Local Zenith and True North

U.S. Naval Obs., Washington
Location: W 77°04′00.0″, N38°55′18.0″, 92m
(Longitude referred to Greenwich meridian)

Astronomical Almanac Bright Star List

Object ID	Date	Time	Zenith Distance	Azimuth (E of N)	V	B-V	U-B	Notes	Spectral Type
	(UT1)								
		h m s	° ′ ″	° ′ ″	mag	mag	mag		
CAPELLA	2004 Sep 01	00:00:00.0	94 59 37.7	2 54 33.2	0.08	0.80	0.44	hcd67	G6 III + G2 III

3.4.1.3 Precision and Accuracy

Except for the topocentric horizon calculations, positions are displayed to the following precision:

- rectangular coordinates and distances: 10^{-9} astronomical unit (AU) or, for the geocentric Moon, 10^{-3} km
- equatorial spherical coordinates: 10^{-3} second of time in right ascension, 0.01 arcsecond in declination
- ecliptic spherical coordinates: 0.1 arcsecond in both ecliptic longitude and latitude
- the equation of time (see this note on the Apparent Geocentric Equator of Date position type in Table 3.1) is given to a precision of 0.1 of a second of time.

MICA's position calculations are based on standard algorithms and are carried out to a precision of better than one milliarcsecond; that is, to at least an order of magnitude better than the tabulation. However, do not confuse precision with accuracy. Note the following:

- In many cases the external accuracy of the calculated positions is limited by the quality of the basic reference data—the fundamental star catalogs and planetary ephemerides. Inadequacies in the current standard models for precession and nutation are also known which affect some position types. A full discussion of this subject is quite complex and cannot be given here; but, in practice, the uncertainties in MICA's tabulated angular coordinates will usually fall between 0.01 and 1 arcsecond, depending on object, position type, and date.
- The positions of solar system bodies are those of their centers of mass, which are not directly observable. However, except for the Moon, the angular coordinates of the center of mass can be assumed to be at the geometric center of the visible (but fully illuminated) disk to the accuracy of the tables.
- For all position types in which the selected time scale is UT1, the uncertainty in the Delta-T extrapolation limits the accuracy of the coordinates of fast-moving solar system objects. For example, the Moon's geocentric angular coordinates, as a function of UT1, may be uncertain at a level of a few tenths of an arcsecond.
- Topocentric apparent horizon coordinates are given to a precision of 0.1 arcsecond in both zenith distance and azimuth. However, when the selected time scale is UT1, the accuracy is limited by the neglect of polar motion—generally, the uncertainty will be of order 0.5 arcsecond. (In practice, the conversion of local clock time to UT1 could be more problematic.) In addition, atmospheric refraction is not taken into account. Refraction can affect zenith distance by several arcminutes (more near the horizon) at optical wavelengths.
- The positions of components of double or multiple stars do not include the effect of orbital motion except that (linear) part included in the position and proper motion in the star catalog; in some cases, catalogs give data on the center of mass or the center of light.

3.4.2 Date & Time

The Date & Time section of MICA 2.0 generates some of the time and reference system values that are tabulated in Section B and section K of The Astronomical Almanac. Select Date & Time from the MICA Calculate/Compute menu. The Date & Time sub-menu contains five items:

- Sidereal Time
- Nutation and Obliquity
- Calendar
- Equation of the Equinoxes
- Delta T

For each of these Date & Time functions, enter the date and time for the computation, the number of repetitions, and the location (for sidereal time only).

3.4.2.1 Sidereal Time

The MICA Sidereal Time menu option tabulates the Greenwich mean and apparent sidereal time, the local mean and apparent sidereal time, and the equation of the equinoxes for a given date. Sidereal time is equal to the hour angle of the catalog equinox and is a direct measure of the diurnal rotation of the Earth. Apparent sidereal time is defined by the intersection of the true equator of date with the ecliptic of date and is affected by the precession and nutation of the axis of the Earth. Mean sidereal time is measured against the mean equinox of date and is only affected by precession. The equation of the equinoxes is equal to the apparent sidereal time minus the mean sidereal time.

The sidereal time output table depends upon the selected Time Scale (TT or UT1). The time scale in use (TT or UT1) is given in the table heading over the time argument. If the UT1 time scale is selected (the usual choice for this calculation), the mean and apparent Greenwich sidereal times will be tabulated, and it is assumed that the observer's longitude is referred to the Greenwich meridian. If the Terrestrial Time scale (TT) is selected, then mean and apparent ephemeris (or dynamical) sidereal times will be tabulated, and it is assumed that the observer's longitude is referred to the ephemeris meridian. The table headings always indicate which quantities are being tabulated and to which prime meridian the longitude of the observer is referred. Note that no time zone correction is applied to the 'Local Sidereal Time' output. The Local Sidereal Time is given in the selected Time Scale (UT1 or TT). Zone time is only used in Rise/Set and Sky Map.

MICA Sidereal Time Output Table:

SIDEREAL TIME

U.S. Naval Obs., Washington
Location: W 77°04'00.0", N38°55'18.0", 92m
(Longitude referred to Greenwich meridian)

Date	Time (UT1)	Greenwich Sidereal Time Mean	Greenwich Sidereal Time App.	Local Sidereal Time Mean	Local Sidereal Time App.	Equation of the Equinoxes
	h m s	h m s	s	h m s	s	s
2003 Oct 08	00:00:00.0	1 04 52.4571	51.6014	19 56 36.4571	35.6014	-0.8557

Regardless of the time scale, all of the tabulated sidereal times are given in hours, minutes and seconds—rounded to the nearest 10^{-4} second. The hours and minutes refer to both the mean and apparent sidereal times, since the two quantities differ by very little. The tabulated seconds may occasionally be greater than or equal to 60. This indicates that the associated sidereal time lies in the next whole minute greater than the one given in the table. For example, a tabulated apparent sidereal time of 6h 40m $60^{\mathrm{s}}.1000$ is equivalent to 6h 41m $00^{\mathrm{s}}.1000$. The equation of the equinoxes is also tabulated to the nearest 10^{-4} second of time.[1]

[1] **Note:** The equation of time is not tabulated here; it is tabulated along with the apparent geocentric coordinates of the Sun in the MICA Positions section.

3.4.2.2 Nutation and Obliquity

MICA Nutation and Obliquity calculations refer to the orientation of the Earth's rotation axis with respect to its orbital plane and are independent of the observer's location. This section of MICA tabulates the date, time, mean and true obliquity of the ecliptic, and nutation in longitude and obliquity. The obliquity of the ecliptic is the angle between the planes of the equator and the ecliptic. The mean obliquity of the ecliptic is the angle between the mean ecliptic and mean equator of date, while the true obliquity of the ecliptic is the angle between the mean ecliptic of date and the true equator of date. The equation for the mean obliquity is given by

$$\varepsilon_0 = 23°26'21.448'' - 46.8150''*T - 0.00059''*T^2 + 0.001813''*T^3$$

where

$$T = (JD - 2451545.0)/36525 \quad \text{(the time in Julian centuries).}$$

Nutation may be resolved into two components; nutation in longitude and nutation in obliquity. These quantities are derived directly from an implementation of the complete 1980 IAU Theory of Nutation and contain no other corrections. See Section 3.222 in the *Explanatory Supplement to The Astronomical Almanac (Seidelmann, 1992)* for further information.

An example of the MICA Nutation and Obliquity output table is shown below. The mean and true obliquity of the ecliptic (Obliq. of Ecliptic) are tabulated in degrees, minutes, and seconds—rounded to the nearest 10^{-3} arcsecond. The degrees and minutes refer to both the mean and true obliquities, since the two quantities differ by very little. The nutation components in longitude (Long.) and obliquity (Obliq.) are tabulated to the nearest 10^{-3} arcsecond.

MICA Nutation and Obliquity Output Table:

```
                         NUTATION AND OBLIQUITY

   Date          Time          Obliq. of Ecliptic          Nutation in
          (UT1)                         Mean    True      Long.     Obliq.
                    h  m   s    °  '      "       "         "          "
2003 Oct 08 00:00:00.0       23 26  19.685 25.936       -13.989    + 6.251
```

3.4.2.3 Calendar

The MICA Calendar menu option shows the correspondence between Julian dates and civil calendar dates. It also gives the day of the week and the day of the year. In the Julian calendar a common year is defined to comprise 365 days, and every fourth year is a leap year comprising 366 days (with a few exceptions). The Julian date is the interval of time in days and fraction of a day since 4713 B.C.E. January 1, Greenwich noon, on the Julian calendar. The Julian Date is often used in precise astronomical studies. Note that the civil calendar is based on the Gregorian calendar.

Since the Calendar Output Table contains both Julian and calendar dates, its format does not depend on the Date System selected. The calendar date appears first, in year, month, day, hour, minute, and seconds format. Next, a three-letter abbreviation specifies the day of the week. Then follows the Julian date rounded to the nearest 10^{-6} day (roughly a tenth of a second of time). The final tabulated quantity is the day of the year, again rounded to the nearest 10^{-6} day.

MICA Calendar Output Table

CALENDAR

Date	Time	Day	Julian Date	Day-of-Year
(UT1)			(UT1)	(UT1)
d	h m s		d	d
2003 Oct 08	00:00:00.0	Wed	2452920.500000	281.000000

Consistent with the other output tables, the table heading indicates the selected time scale. Note, however that the Calendar selection does not perform a conversion between time scales. All quantities tabulated are assumed to be in the same time scale.

3.4.2.4 Equation of the Equinoxes

The MICA Equation of the Equinoxes menu option tabulates the calendar date and time, Julian date, and the equation of the equinoxes. The equation of the equinoxes is equal to the apparent sidereal time minus the mean sidereal time. It is affected by the motion of the equinox due to nutation and is tabulated to the nearest 10^{-4} second of time. See the Date & Time Sidereal Time menu option for further information on the equation of the equinoxes.

Equation of the Equinoxes

EQUATION OF THE EQUINOXES

Date	Time	Julian Date	Equation of the Equinoxes
(UT1)			
	h m s	d	s
2003 Oct 08	00:00:00.0	2452920.500000	-0.8557

3.4.2.5 Delta T

The MICA Delta T menu option tabulates the calendar date and time, day of the week, Julian date, and the Delta T value. Delta T (ΔT) is the difference between Terrestrial Time (TT) and Universal Time (UT): $\Delta T = TT - UT1$. Variations in Delta T are caused by changes in the rotational speed of the Earth. MICA 2.0 utilizes an internal table of historical and predicted Delta T values. The Delta T value determined by MICA on a specific date is derived by a linear interpolation between the adjacent Delta T table entries. For dates before 1972, the historical values of Delta T were derived from McCarthy and Babcock.[1] Delta T values after 1972 to the present (or date the Delta T table was updated) were derived from the 'finals.data' and the 'cumulative number of leap seconds' file on the USNO IERS *Bulletin A* web site. Predicted values of Delta T to 2050 were supplied by Johnson

[1] *Physics of the Earth and Planetary Interiors*, Vol. 44, 1986, p. 281.

(2004). Predicted Delta T errors range from about $\pm 0.004^s$ in early 2005, to $\pm 24^s$ in 2050. Updates to the Delta T table will be periodically released on the USNO MICA web site in order to improve the accuracy of the Delta T computations.[1]

Delta T

DELTA T: TT-UT1

Date	Time (UT1)	Day	Julian Date (UT1)	Delta T
	d h m s		d	s
2003 Oct 08	00:00:00.0	Wed	2452920.500000	64.554

3.4.3 Rise/Set/Transit

The Rise/Set/Transit section of MICA generates rise and set times for the Sun, Moon, major planets, selected asteroids, and selected bright stars. It also has the capability to do rise/set/transit calculations for stellar or star-like objects listed in external catalogs. Times of twilight are also tabulated for the Sun.

To obtain a tabular listing of rise/set/transit times on the PC, select Rise/Set/Transit from the MICA Calculate menu. If an asteroid, bright star, or a catalog object is wanted, the desired object(s) should be specified in the respective asteroid, bright star, or catalog dialog boxes. Next, enter the date and time, number of repetitions of the calculation, the Time System, and observer's location in the Date/Time/Location dialog box. Rise/Set/Transit calculations may be done utilizing the TT or UT1 time scales or for a specific local time zone. The Time Zone is entered via the 'Change?' button. The Twilight type (for the Sun only) is entered via the 'More Options' button.

To obtain a tabular listing of rise/set/transit times on the Mac, select Rise/Set/Transit from the MICA Compute menu and then select the desired object from the displayed list. Next a Rise/Set dialog box will be displayed, with separate views for entry of the Date, number of Repetitions, Location, and Time Zone information. If a rise/set calculation for the Sun was selected, then an additional view is present for selection of the type of Twilight calculation. If a rise/set calculation for an asteroid or a bright star is selected, then an additional view is present for selection of the specific object. If a catalog rise/set calculation is selected, then two extra views (Catalog and Search) are present, one for selection of the catalog and the second to specify the catalog search criteria.

3.4.3.1 Rise and Set

The time of rising and setting of a celestial body is defined to occur when the upper limb of the body, affected by refraction, appears on the horizon of an observer. In MICA 2.0, the user can specify the height of the observer. The height can range from the surface of the Earth to a maximum of 11 km (the top of the troposphere). It is also assumed that the observer's horizon is flat (i.e., no geometric 'dip' is applied) and unobstructed.

[1] http://aa.usno.navy.mil/software/mica/delta_T_updates.html

MICA tabulates rise and set times to a precision of one minute only (i.e., no seconds are tabulated). This is because the observed times of rise and set are affected by random changes in local atmospheric conditions and other local variables which cannot be accurately modeled. Thus, tabulating the times to a higher precision is not practical or normally useful.

There are several differences between the algorithms used to determine times of rise and set in MICA 1.0 and MICA 2.0. MICA 1.0 assumed that the observer was at the surface of the Earth (i.e., the height above the Earth was ignored). MICA 2.0 allows any height up to 11 km. MICA 1.0 also assumed that the amount of refraction at the horizon was a standard 34'. MICA 2.0 incorporates a model that determines the angular refraction by numerically integrating a ray passing through a simple polytropic atmosphere. The model is based on the method described by Hohenkerk and Sinclair (1985) and by Hohenkerk, et al. (1992). Lastly, the MICA 2.0 rise/set algorithm works by interpolation, not iteration. One advantage of the interpolation method is that it is more likely to accurately detect and predict rise/set phenomena at high latitudes (i.e., latitudes above about ±65°).

For rise and set calculations, MICA allows the user to specify a time zone and standard or daylight saving time, so that the tabulated rise and set times will be expressed in local time. Actual local time is based on UTC, while MICA's local times are based on UT1. The difference is always less than 0.9 second, which is not significant for rise and set times given the other uncertainties.

3.4.3.2 Twilight

MICA automatically tabulates the times of the beginning and end of twilight when the times of sunrise and sunset are requested. MICA offers a choice of three different periods of twilight.

- **Civil twilight** begins in the morning and ends in the evening when the geometric center of the Sun is 6° below the horizon. Before the beginning of civil twilight in the morning and after its end in the evening, artificial illumination is normally required to carry on ordinary outdoor activities.
- **Nautical twilight** begins and ends when the geometric center of the Sun is 12° below the horizon. Under ordinary atmospheric conditions, only general outlines of ground objects are distinguishable during nautical twilight; for navigational purposes, the horizon is not ordinarily visible before the beginning of morning nautical twilight and after the end of evening nautical twilight.
- **Astronomical twilight** begins and ends when the geometric center of the Sun is 18° below the horizon. From the end of evening astronomical twilight to the beginning of morning astronomical twilight, the sun does not contribute to sky illumination.

Although the mathematical definitions of 6°, 12°, and 18° are precise, the rules of thumb describing illumination during the twilight periods assume clear skies and are subject to local weather, lighting conditions, and altitude.

3.4.3.2.1 Rise/Set/Transit Output Tables

MICA's rise and set calculations produce a tabulation of local rise, set, and transit times, the local azimuths at rise and set (measured eastward from true north) and the local altitude at transit. The times of rise and set correspond to the times when the upper limb of the object appears to be coincident with a level, unobstructed horizon. The computed altitude at transit is topocentric, for the center of the disk, and includes the effect of refraction. For the Sun, the times of the beginning and end of twilight are also shown. The twilight conditions are defined above. All times are tabulated to the nearest minute, and all angular measures are tabulated to the nearest degree. Local topography, elevation of the observer, and meteorological conditions may cause the observed rise and set times to differ from what is tabulated.

Rise/Set/Transit Table for the Sun

```
                                   Sun

                        U.S. Naval Obs., Washington
                Location:  W 77°04'00.0", N38°55'18.0",     92m
                  (Longitude referred to Greenwich meridian)

                   Time Zone:  5h 00m west of Greenwich

      Date             Begin      Rise   Az.    Transit Alt.    Set   Az.      End
     (Zone)            Civil                                                   Civil
                      Twilight                                                Twilight
                        h  m       h  m    °      h  m   °       h  m    °       h  m
2003 Dec  1 (Mon)     06:38      07:08 118      11:57 30S      16:47 242      17:16
2003 Dec  2 (Tue)     06:39      07:09 118      11:58 30S      16:46 242      17:16
```

The date is given on the left side of each line of the table for the Sun, Moon, major planets, asteroids and selected bright stars. The object name, followed by the date, is given on the left side of each line of the table for catalog objects. Each rise/set/transit phenomena that occurs on that date is listed on this line. Note that corresponding rise, set, transit, and/or twilight times may not necessarily occur on the same date, but may occur on the date before or the date after the requested date. Occasionally, an object may experience a second rise/set/transit phenomena on a given date (depending on the right ascension and its rate of change). MICA recognizes these cases. The second phenomena will be shown on a second output line for that date.

Circumpolar objects experience both an upper and a lower transit event in a single day. The extra lower transit event is shown on the right side of the rise/set/transit table. An upper and lower transit may also occur when the observer is at a high latitude.

Blanks may occur in a rise/set table for the Sun, Moon or other objects. This indicates that this particular rise or set event did not occur on the given day. These blanks may occur for a couple of reasons. Blanks occur in the tables in high latitude situations where the object may rise and then be continuously above the horizon for an extended period of time, or conversely finally set after being above the horizon for an extended period of time. Blanks may also occur in a rise/set table as the time

Rise/Set/Transit Table for Object with an Upper and Lower Transit

Hadar

Location: W 77°04'00.0", S38°55'18.0", 92m
(Longitude referred to Greenwich meridian)

Date (UT1)	Rise h m	Az. °	Transit h m	Alt. °	Set h m	Az. °	Lower Transit h m	Al °
2003 Dec 30 (Tue)	*****	***	12:38	69S	*****	***	00:40	10
2003 Dec 31 (Wed)	*****	***	12:34	69S	*****	***	00:36	10
2004 Jan 01 (Thu)	*****	***	12:30	69S	*****	***	00:32	10
2004 Jan 02 (Fri)	*****	***	12:26	69S	*****	***	00:28	10

Rise/Set/Transit calculation for the MASC catalog object, Altair
Note there are additional catalog fields on the right side of the table which are not displayed below

U.S. Naval Obs., Washington
Location: W 77°04'00.0", N38°55'18.0", 92m
(Longitude referred to Greenwich meridian)

MICA Astrometric Stellar Catalog

Object ID	Date (UT1)	Rise h m	Az. °	Transit h m	Alt. °	
ALTAIR	2003 Dec 30 (Tue)	11:53	78	18:24	61S	0

Table 3.2 Rise/Set Table Symbols

Symbol	Definition
*****	There is no event because the object is continually above the horizon.
----	There is no event because the Sun is continually below the horizon.
N	Altitude at local transit is measured from the northern horizon.
S	Altitude at local transit is measured from the southern horizon.
?????	Phenomenon is indeterminate.

Table 3.3 Twilight Table Symbols

Symbol	Definition
/////	There is no event because the Sun is continually above the twilight zenith distance.
----	There is no event because the object is continually below the twilight zenith distance.

of rise/set changes across a day boundary (e.g Sun rises later and later in the day as summer approaches). Blanks may also occur in the moonrise/moonset table because the time between successive moonrises or moonsets is about 25 hours or about one hour longer than the 24 hour day. Consequently, these gaps in the moonrise/moonset table occur approximately once every 25 days (or so).

For catalog objects, several additional catalog-dependent fields are included in the rise/set/transit catalogue table output. Table 3.2 lists the symbols that may appear in these tables. For Sun calculations for high latitudes, one may also see the symbols shown in Table 3.3 in the twilight columns.

3.4.4 Physical Ephemeris

MICA generates physical ephemerides for the Sun, Moon, and major planets. Physical ephemerides describe the appearance of the solar, lunar and planetary disks as seen from a point on or near the Earth. Both illumination and rotation parameters are available for all listed bodies, except for the Sun.

To obtain a tabular listing, for the PC, select Physical Ephemeris from the MICA Calculate menu. Next select the desired Object (Sun, Moon, or major planets), Origin (Topocentric or Geocentric), Physical Ephemeris Type (Rotation or Illumination), and Rotation System (for Jupiter only). On the next dialog box, enter the date and time, number of repetitions of the calculation, the time system, and observer's location (for topocentric calculations only).

To obtain a tabular listing, for the Mac, select either the Illumination of Disc or Rotation of Disc from the MICA Compute menu and then select the desired object from the displayed list. Next, a physical ephemeris dialog box will be displayed with separate views for entry of the origin Type (Geocentric or Topocentric) and Rotation System (Jupiter only), Date/Time, number of Repetitions, and Location (topocentric only).

3.4.4.1 Physical Ephemerides Calculations

The physical ephemeris of a solar system body refers to its aspect as seen from the Earth: its apparent magnitude, the angular size of its disk, its apparent degree of illumination, the orientation of its pole, and the positions of its sub-solar and sub-Earth points. This information divides into illumination data and rotation data. Illumination data depend on a model of the body's reflectivity as a function of angle of illumination, while rotation data depend on a model of the body's rotation; both depend on the Earth-Sun-body geometry. Illumination information includes the object's apparent phase, magnitude, and the angular dimensions of its disk. Rotation information includes the instantaneous positions of the sub- solar and sub-Earth points on the object's surface and the apparent position angle of the object's axis of rotation.

In MICA's physical ephemeris tabulations, longitudes and latitudes are planetographic, and position angles are measured on the sky eastward from true north (the direction to the true, Celestial Ephemeris Pole of date). The illumination and rotation data are the same as those found in the 'Ephemeris for Physical observations' section in Section E of *The Astronomical Almanac*. All of MICA's physical ephemerides are available using either a geocentric or topocentric origin. Except for the Moon, and Venus and Mars when near the Earth, the geocentric physical ephemeris of an object is indistinguishable from the topocentric physical ephemeris. Rotation data are available for the Sun, Moon and major planets. Illumination data are available for the Moon and major planets, but not the Sun.

The MICA physical ephemerides of the planets have been calculated using the basic physical data (directions of the north poles of rotation, the prime meridians, and the size and shapes of the major planets) contained in Seidelmann (2002). Expressions for the apparent visual magnitudes of the major planets are from Harris (1961). Values for V(1,0), the magnitude of the planet as seen from 1

AU and at a phase angle of 0°, are given on page E4 of *The Astronomical Almanac*. The MICA 2.0 expressions for the magnitudes of Mercury and Venus are based on the parameters given in Hilton (2003) and are the same as used in the 2005 and 2006 editions of *The Astronomical Almanac*. The MICA 2.0 Mercury and Venus magnitudes differ slightly from the 2004 edition (and earlier editions) of *The Astronomical Almanac*, which used the earlier Harris (1961) expressions. A useful discussion on the calculation of physical ephemerides is also contained in Hilton (1992).

A number of examples of physical ephemeris output tables are shown below. Since the quantities listed in the output table will depend strongly on your menu selections, the various forms of the tables are described separately below.

3.4.4.2 Rotation Parameters

Rotation data are available for the Sun, Moon and major planets.

3.4.4.2.1 Rotation Output Tables—Planets

Tabulated longitudes and latitudes of the sub-Earth and sub-solar points are in planetographic coordinates. Planetographic longitude (Long.) is reckoned from the prime meridian and increases from 0° – 360° in the direction opposite rotation. The planetographic latitude (Lat.) of a point is the angle between the planet's equator and the normal to the reference spheroid of the planet at the point. Latitudes north of the equator are positive. For Jupiter, three longitude systems are defined, each system corresponding to a different apparent rate of rotation. System I represents the rotation of the visible cloud layer at low latitudes, System II represents the rotation of the visible cloud layer at high latitudes, and System III represents an internal rotational system (possibly that of the core) derived from radio observations. In MICA, the rotational system is selectable for Jupiter only; only Jupiter has all three systems currently defined. For the other giant planets, only the System III rotations are now used.

Physical Ephemeris Rotation Parameters

Jupiter

Geocentric Physical Ephemeris: Rotation Parameters
Rotation System I

Date	Time (UT1)	Sub-Earth Pt. Long.	Sub-Earth Pt. Lat.	Sub-Solar Point Long.	Sub-Solar Point Lat.	Sub-Solar Point Dist.	Sub-Solar Point P.A.	North Pole Dist.	North Pole P.A.
	h m s	°	°	°	°	"	°	"	°
2003 Oct 08	00:00:00.0	167.52	-1.06	173.72	-0.94	1.72	113.31	-14.95	24.28
2003 Oct 09	00:00:00.0	325.27	-1.08	331.58	-0.94	1.76	113.30	-14.97	24.30
2003 Oct 10	00:00:00.0	123.03	-1.09	129.46	-0.95	1.79	113.30	-14.99	24.33
2003 Oct 11	00:00:00.0	280.78	-1.10	287.33	-0.95	1.83	113.29	-15.02	24.35
2003 Oct 12	00:00:00.0	78.54	-1.11	85.20	-0.96	1.86	113.29	-15.04	24.38

At the center of the apparent disk of a planet is the sub-Earth point; other reference points are the sub-solar point and the north pole. For an oblate planet, the Earth and Sun are not at the zeniths of the sub-Earth and sub-solar points, respectively. The apparent distance (Dist.) and the position angle (P.A.) with respect to the center of the disk are given for both the sub-solar point and the north pole. Position angles are measured east from the north on the celestial sphere, with north defined by the great circle on the celestial sphere passing through the center of the planet's apparent disk and the true celestial pole of date. The apparent distance is given in arcseconds. Tabulated distances are positive for points on the visible hemisphere of the planet and negative for points on the far side. Thus, as the north pole or sub-solar point passes from the visible hemisphere to the far side, or vice versa, the sign of the distance changes abruptly; but the position angle varies continuously. However, when the point passes close to the center of the disk, the sign of the distance remains unchanged; but both distance and position angle vary rapidly.

All tabulated quantities are corrected for light-time, so the given values apply to the disk that is visible at the tabulated time. Except for planetographic longitudes, all tabulated quantities vary so slowly that they remain unchanged if the time scale used is UT1 or TT. A change of time scale from TT to UT1 affects the tabulated planetographic longitudes by several tenths of a degree for all the planets except Mercury, Venus, and Pluto.

3.4.4.2.2 Rotation Output Tables—Moon

The positions of points on the Moon's surface are specified by a system of selenographic coordinates, in which the latitude (Lat.) is measured positively to the north from the equator of the pole of rotation, and longitude (Long.) is measured positively to the east on the selenocentric celestial sphere from the lunar meridian through the mean center of the apparent disk. Selenographic longitudes are measured positive to the west (towards Mare Crisium) on the apparent disk; this sign convention implies that the longitudes of the Sun and of the terminators are decreasing functions of time.

Moon

Geocentric Physical Ephemeris: Rotation Parameters

Date	Time	Selenographic Coordinates				Position Angles	
		Earth		Sun		Axis	Bright
(UT1)		Lat.	Long.	Lat.	Long.		Limb
	h m s	°	°	°	°	°	°
2003 Oct 08	00:00:00.0	6.030	5.559	0.97	31.81	337.500	237.58
2003 Oct 09	00:00:00.0	5.232	4.925	0.95	19.65	337.495	231.76
2003 Oct 10	00:00:00.0	4.200	4.081	0.93	7.49	338.375	204.67
2003 Oct 11	00:00:00.0	2.990	3.043	0.91	355.33	340.058	85.00
2003 Oct 12	00:00:00.0	1.660	1.839	0.89	343.17	342.492	74.64

The position angle of the axis of rotation and the midpoint of the bright limb are measured counterclockwise from the north. The position angle of the terminator differs by 90° from the position angle of the bright limb.

On the average, the same hemisphere of the Moon is always turned towards the Earth, but there is a periodic oscillation or libration of the apparent position of the lunar surface that allows about 59% of the surface of the Moon to be seen from the Earth. The libration is due, partly, to a physical libration, which is an oscillation of the actual rotational motion about its mean rotation (analogous to the

Earth's nutation). The main contributor to libration, however, is the much larger geocentric optical libration (a better term is probably geometric libration), which results from the eccentricity and inclination of the Moon's orbit about the Earth. Both of these effects are taken into account in the computation of the Earth's geocentric selenographic longitude and latitude and the position angle of the axis of rotation. There is a further contribution to the optical libration due to the difference between the viewpoints of the observer on the surface of the Earth and the hypothetical observer at the geocenter. These topocentric optical librations may be as much as 1° and should be noticeable in the output tables if the origin is changed from Geocentric to Topocentric.

When the libration in longitude, i.e., the selenographic longitude of the Earth, is positive, the point at the mean center of the disk is displaced eastward on the celestial sphere, exposing to view a region of the west limb. When the libration in latitude, or selenographic latitude of the Earth, is positive, the point at the mean center of the disk is displaced towards the south, and a region of the north limb of the Moon is exposed to view. In a similar way the selenographic coordinates of the Sun show which regions of the lunar surface are illuminated.

The selenographic coordinates of the Earth and the Sun specify the points on the lunar surface where the Earth and Sun, respectively, are in the selenographic zenith. The selenographic longitude and latitude of the Earth take into account the total optical and physical librations in longitude and latitude, respectively. When a Topocentric origin has been selected, the 'Earth' should be understood to mean the specific location of the observer. New Moon, First Quarter, Full Moon, and Last Quarter correspond, approximately, to a selenographic longitude of the Sun of 180°, 90°, 0°, and 270°, respectively. As viewed from the Earth, lunar longitudes are measured positively towards the west (towards Mare Crisium). This sign convention implies that the longitudes of the sub-solar point and the terminator are decreasing functions of time. *The Astronomical Almanac* tabulates the colongitude of the Sun which is 90° (or 450°) minus its longitude, and increasing function of time

3.4.4.2.3 Rotation Output Tables—Sun

The position angle of the pole of the Sun and the heliographic latitude (Lat.) and longitude (Long.) of the Earth are from Carrington (1863). The horizontal parallax of the Sun is calculated as the inverse sine of the Earth's equatorial radius divided by the distance of the Sun's center from the geocenter. The apparent semidiameter of the Sun is the inverse sine of the Sun's radius divided by the distance of the Sun's center from the selected origin (Geocentric or Topocentric).

3.4.4.3 Illumination Parameters

3.4.4.3.1 Illumination Output Tables—Planets

The tabulated light-time, in minutes, is the travel time for light arriving at the Earth at the tabular time. When Mercury and Venus are near conjunction with the

Sun

Geocentric Physical Ephemeris: Rotation Parameters

Date	Time (UT1)	Position Angle of Axis	Heliographic Lat.	Heliographic Long.	Horizontal Parallax	Semi-Diameter
	h m s	°	°	°	"	' "
2003 Oct 08	00:00:00.0	26.26	6.37	205.03	8.80	16 00.30
2003 Oct 09	00:00:00.0	26.28	6.31	191.84	8.80	16 00.58
2003 Oct 10	00:00:00.0	26.29	6.25	178.65	8.81	16 00.86
2003 Oct 11	00:00:00.0	26.29	6.18	165.45	8.81	16 01.13
2003 Oct 12	00:00:00.0	26.29	6.12	152.26	8.81	16 01.41

Sun, the equations that describe their magnitudes (Mag.) are outside their valid ranges and may be unreliable. Therefore, under these circumstances, the magnitudes for the inferior planets are not tabulated. For Saturn, the magnitude includes the contribution due to the rings.

Jupiter

Geocentric Physical Ephemeris: Illumination Parameters

Date	Time (UT1)	Light-time	Mag.	Diameter Equator	Diameter Pole	Phase	Phase Angle	Sun's Plan. Orb. Long.
	h m s	min.		"	"		°	°
2003 Oct 08	00:00:00.0	51.28	-1.8	31.97	29.90	0.997	6.2	195.24
2003 Oct 09	00:00:00.0	51.20	-1.8	32.02	29.94	0.997	6.3	195.32
2003 Oct 10	00:00:00.0	51.12	-1.8	32.07	29.99	0.997	6.4	195.39
2003 Oct 11	00:00:00.0	51.04	-1.8	32.12	30.04	0.997	6.5	195.47
2003 Oct 12	00:00:00.0	50.96	-1.8	32.17	30.09	0.997	6.7	195.55

Unlike *The Astronomical Almanac*, MICA does not calculate the surface brightness of the planets because the surface brightness is fairly constant and variations resulting from surface markings, clouds, and atmospheric light scattering cannot be taken into account without using highly sophisticated models of the planet's surface and atmosphere. Even these models will miss transient effects (storms of various types) which can be significant. For those who want to compute an approximate, mean surface brightness (BS) of the planet, in magnitudes per square arcsecond, the formula is:

$$B_S = V + 2.5 \log_{10}(k\pi ab)$$

where V is the visual magnitude of the planet, a and b are the apparent equatorial and polar semidiameters of the planet (in arcseconds), and k is the fraction of the disk illuminated (phase of the planet). MICA also does not calculate the greatest defect of illumination; that is, the longest apparent angular length of the observed lunar or planetary disk that is not illuminated to an observer on the Earth. The length of the greatest defect of illumination (q) to first order is:

$$q = 2a(1-k)[1-(1-b/a)\sin^2(PA_s - PA_n + 90°)]$$

where PA_s is the position angle of the sub-solar point and PA_n is the position angle of the north pole of the planet. For a spherical planet, this formula reduces to $q = 2a(1-k)$. The position angle of the greatest defect of illumination is $PA_s + 180°$.

The apparent disk of an oblate planet is always an ellipse, with an oblateness less than or equal to the oblateness of the planet itself, depending on the apparent tilt of the planet's axis. Separate apparent equatorial and polar diameters are tabulated for all of the planets.

The apparent phase is the ratio of the apparent illuminated area of the disk to the total area of the disk, as seen from the Earth. The phase angle is the planetocentric elongation of the Earth from the Sun. Calculations of the phase and phase angle are based on the geometric terminator, which is defined by the plane crossing through the planet's center of mass, orthogonal to the direction of the Sun.

The planetographic orbital longitude of the Sun (labelled Sun's Plan. Orb. Long.) is measured eastward in the planet's orbital plane from the planet's vernal equinox. Instantaneous orbital and equatorial planes are used in computing the planetographic orbital longitude of the Sun. Values of 0°, 90°, 180° and 270° correspond to the beginning of spring, summer, autumn and winter, respectively, for the planet's northern hemisphere.

3.4.4.3.2 Illumination Output Tables—Moon

Just as for the planets, the phase (Fraction Illuminated) of the Moon is the ratio of the apparent illuminated area of the disk to the total area of the disk, as seen from the Earth. The phase angle is the selenocentric elongation of the Earth from the Sun. Calculations of the phase are based on the geometric terminator, which is defined by the plane crossing through the Moon's center of mass, orthogonal to the direction of the Sun.

Moon

Geocentric Physical Ephemeris: Illumination Parameters

Date	Time (UT1) h m s	Phase Angle °	Fraction Illuminated	Horizontal Parallax ' "	Semi-diameter ' "
2003 Oct 08	00:00:00.0	26.68	0.947	55 43.29	15 10.67
2003 Oct 09	00:00:00.0	15.31	0.982	55 17.00	15 03.51
2003 Oct 10	00:00:00.0	4.72	0.998	54 53.64	14 57.15
2003 Oct 11	00:00:00.0	7.98	0.995	54 33.68	14 51.71
2003 Oct 12	00:00:00.0	18.68	0.974	54 17.90	14 47.42

The horizontal parallax of the Moon is calculated as the inverse sine of the Earth's equatorial radius divided by the distance of the Moon's center from the geocenter. The apparent semidiameter of the Moon is the inverse sine of the Moon's radius divided by the distance of the Moon's center from the selected origin (Geocentric or Topocentric).

3.4.5 Configurations

The Configurations section of MICA provides tables of low precision topocentric position data for the Sun, Moon, major planets, and selected asteroids at specified time(s). A sky map can also be displayed which shows the topocentric positions of the Sun, Moon, major planets, selected asteroids, and bright stars at a specific time. This section of MICA is designed to provide "quick look" information that should be useful for tasks such as planning an observing session or pointing a telescope at one of the objects.

Select Configurations from the MICA 2.0 Calculate/Compute menu. The Configurations sub-menu contains 3 items:

- Solar System
- Asteroids
- Sky Map

For the PC, next enter the date and time, number of repetitions of the calculation, the Time System (TT, UT1, or Zone Time — for the Sky Map only), and observer's location in the Date/Time/Location dialog box. The time zone must be specified via the 'Change?' or 'More Options' buttons. The sky map colors (background, asteroids/planets and stars) are specified via the 'More Options' button. The solar system objects (Sun, Moon, Planets, and/or Asteroids) and stars (All or a magnitude range) to be displayed are selected via the 'More Options' button or in the next dialog box. When specifying a magnitude range remember the lower the magnitude the brighter the star.

The Mac Sky Map dialog box contains 5 separate views for entry of the Date/Time, Location, Objects to be displayed, Time Zone, and Colors to be plotted. The solar system objects (sun, moon, planets, and/or asteroids) and stars (All or a magnitude range) to be displayed are selected via the Objects view. When specifying a magnitude range remember the lower the magnitude the brighter the star. The sky map colors (sun, moon, planets, asteroids, stars and sky) are specified via the Colors view.

3.4.5.1 Solar System Configurations

The heading of the Solar System Configuration table contains the observer's coordinates and the date and time to which the tabular data refers. The body of the table contains one line for each object in the solar system other than the Earth. Separate, complete tables are produced for each tabular date.

Each line in the table lists the following information. The *right ascension* (R.A.) and *declination* (Dec.) columns provide the topocentric apparent equatorial coordinates of each object, rounded to the nearest tenth of a minute of time in right ascension and to the nearest arcminute in declination. The *distance* (Dist.) column gives the true distance from the observer to the object, rounded to the nearest 10^{-3} of an astronomical unit (A.U.) for the Sun and the major planets, and to the nearest kilometer (km) for the Moon. Apparent horizon coordinates, referred to the observer's location, are tabulated in the *zenith distance* (Z.D.) and *azimuth* (Az.) columns. The horizon coordinates are tabulated to the nearest degree (atmospheric refraction is not taken into account). The *elongation* (Elong.) column gives the angular separation between the Sun and the object to the nearest degree. The direction of the object with respect to the Sun is given by the letters N, S, E, or W (north, south, east, or west) immediately preceding the angular measure. The *diameter* (Diam.) column lists the equatorial diameter of the object's apparent disk (fully illuminat-

Solar System Configurations Output Tables

TOPOCENTRIC CONFIGURATION

U.S. Naval Obs., Washington
Location: W 77°04'00.0", N38°55'18.0", 92m
(Longitude referred to Greenwich meridian)

2003 Oct 08 00:00:00.0 (UT1)

Object	R.A.	Dec.	Dist.	Z.D.	Az.	Elong.	Diam.	Mag.
	h m	° '	A.U.	°	°	°	' "	
Sun	12 52.6	- 5 38	0.999	106	276	----	32 00.6	----
Mercury	12 08.9	+ 1 08	1.231	110	289	W 13	0 05.5	-1.3
Venus	13 44.2	- 9 59	1.657	99	264	E 13	0 10.1	-3.7
Mars	22 17.6	-14 55	0.485	63	141	E136	0 19.3	-1.9
Jupiter	10 42.8	+ 9 08	6.166	118	312	W 36	0 32.0	-1.8
Saturn	6 55.9	+22 05	8.952	117	16	W 91	0 18.6	0.1
Uranus	22 06.8	-12 27	19.314	60	143	E135	0 03.6	5.8
Neptune	20 51.7	-17 38	29.623	58	164	E116	0 02.3	7.9
Pluto	17 09.7	-14 00	31.134	66	225	E 64	0 00.1	13.9
	h m	° '	km	°	°	°	' "	Illum.
Moon	23 25.2	- 9 40	391137	69	123	E153	30 32.3	95%

ed), to the nearest tenth of an arcsecond.

The *magnitude* (Mag.) column provides the visual magnitude of each major planet, rounded to the nearest tenth of a magnitude. The magnitudes of the Moon and Sun are not listed. When Mercury and Venus are near conjunction with the Sun, their computed magnitudes are not realistic and are not tabulated (i.e., dashes appear in the table in place of a numerical value). For Saturn, the magnitude includes the contribution due to the rings. For the Moon, the percentage of the disk that is illuminated is given in place of the magnitude.

MICA also checks for the occurrence of eclipses of the Sun and Moon and transits of Mercury and Venus across the Sun. If one of these phenomena occurs, a message will appear directly below the main body of the table. The message will appear only if the phenomenon occurs when the objects involved are above the observer's horizon. It is important to understand that these messages describe the instantaneous status of the phenomenon at the time and location of interest. For example, a message such as "Sun in partial eclipse" means that at the time and location indicated, the Sun is partially obscured by the Moon, not that the eclipse necessarily is classified as a partial eclipse. A configuration table computed for the same date but some other time or location might indicate total or annular obscuration.

3.4.5.2 Asteroid Configurations

The heading of the Asteroid Configuration table contains the observer's coordinates and the date and time to which the tabular data refers. The body of the table contains one line for each individual asteroid. Separate, complete tables are produced for each tabular date.

Each line in the table lists the following information. The astrometric right ascension (R.A.) and declination (Dec.) columns provide the astrometric geocen-

Asteroids Configurations Output Tables

TOPOCENTRIC CONFIGURATION

U.S. Naval Obs., Washington
Location: W 77°04'00.0", N38°55'18.0", 92m
(Longitude referred to Greenwich meridian)

2003 Oct 08 00:00:00.0 (UT1)

	Astrometric		Topocentric						
Object	R.A.	Dec.	R.A.	Dec.	Dist.	Z.D.	Az.	Elong.	Mag.
	h m	° '	h m	° '	A.U.	°	°	°	
Ceres	7 22.1	+23 21	7 22.3	+23 21	2.553	117	9	W 85	8.6
Cybele	0 35.3	+ 0 46	0 35.5	+ 0 47	2.521	74	102	E174	11.6
Davida	11 46.0	+12 42	11 46.2	+12 41	3.830	106	301	W 25	12.0
Eunomia	9 05.2	+16 41	9 05.4	+16 40	2.833	122	340	W 60	10.4
Europa	2 40.0	+ 4 20	2 40.2	+ 4 21	2.093	96	80	W153	10.9
Flora	18 11.3	-24 16	18 11.6	-24 16	2.159	68	206	E 78	10.9
Hebe	7 24.6	+ 8 51	7 24.8	+ 8 51	2.021	132	11	W 83	10.0
Hygiea	7 25.7	+22 43	7 25.9	+22 43	3.372	118	8	W 85	11.6
Interamnia	10 56.0	- 6 10	10 56.2	- 6 11	4.252	128	297	W 29	12.4
Iris	10 03.8	+ 8 14	10 04.0	+ 8 13	2.836	124	321	W 44	10.5
Juno	15 35.2	- 8 39	15 35.4	- 8 39	4.045	77	247	E 40	11.5
Metis	16 38.1	-23 50	16 38.3	-23 51	3.072	78	225	E 57	11.8
Pallas	1 58.4	-15 35	1 58.6	-15 34	1.808	100	102	W153	8.3
Psyche	15 52.3	-17 37	15 52.5	-17 38	3.734	80	238	E 46	12.0
Vesta	15 17.5	-14 53	15 17.7	-14 54	2.869	84	246	E 37	7.9

tric equatorial coordinates of each object, rounded to the nearest tenth of a minute of time in right ascension and to the nearest arcminute in declination. Next, the topocentric right ascension (R.A.) and declination (Dec.) columns provide the topocentric apparent equatorial coordinates of each object. The distance (Dist.) column gives the true distance from the observer to the object, rounded to the nearest 10^{-3} of an astronomical unit (A.U.). Apparent horizon coordinates, referred to the observer's location, are tabulated in the zenith distance (Z.D.) and azimuth (Az.) columns.

3.4.5.3 Sky Map Configurations

The MICA Sky Map window has three major parts. The Sky Map itself is displayed on the left side of the window. The PC version displays the date, time, time scale (UT1 or TT) or time zone, and the observer's latitude and longitude in the upper right corner of the Sky Map. The PC version displays a key or legend identifying the plotted Solar System objects and their altitude and azimuth's are shown on the bottom right side of the screen. The Mac version has a Sky Map dialog box in the upper right side of the map, which can be used to replot the map. The Mac version also displays Selected Object Information on the bottom right side of the screen. The Sky Map is a polar diagram centered on the observer's zenith at the specified date and time. The outer horizon circle is at a zenith distance of 90°. The azimuth directions for North (0°), East (90°), South (180°), and West (270°) are noted just outside the horizon circle.

The topocentric positions (refraction ***not*** included) of all selected objects that are above the horizon at the specified date and time are plotted in the Sky Map on the left. The Sun and Moon are always included if they are above the horizon. The user may select which other objects to include in the plot (major planets, as-

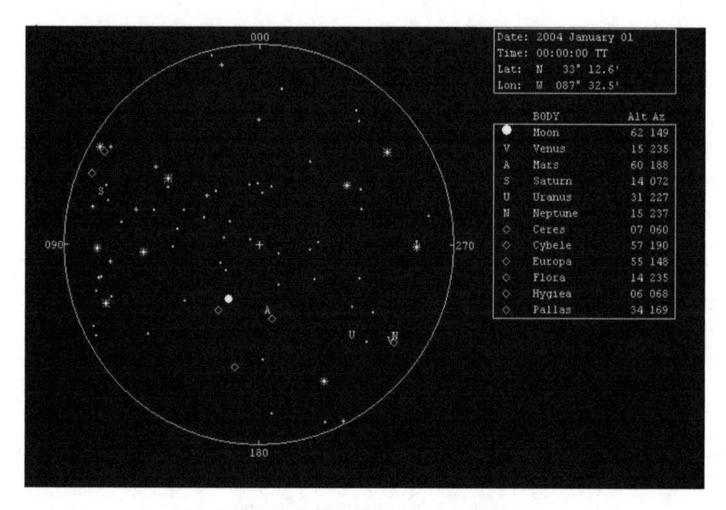

Sky Map configuration for Windows

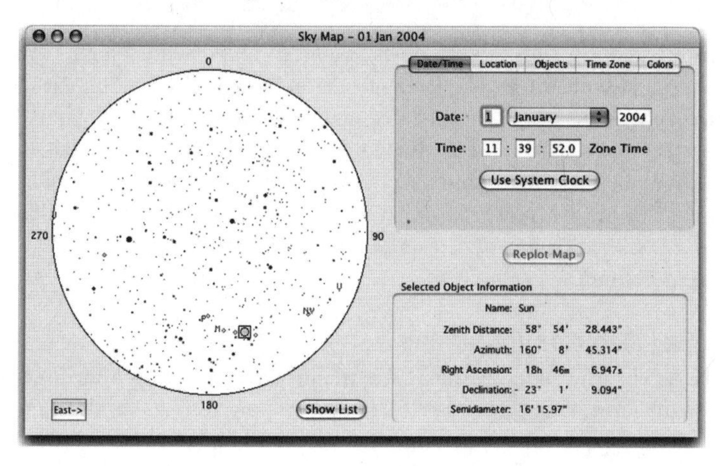

Sky Map configuration for Macintosh

teroids, and/or stars. The Sun is indicated by an open circle with a 'dot' in the center. The Moon is indicated by an filled circle (regardless of phase). Planets are designated by initial letters: M = Mercury, V = Venus, A = Mars (A stands for Ares. M was already taken by Mercury), J = Jupiter, S = Saturn, N = Neptune, U = Uranus, and P = Pluto. The Asteroids are plotted with open diamond symbols. The star data comes from the database of Bright Stars.

The star symbol size and shape for PC s are correlated with brightness:

- stars brighter than 0.5 magnitude use an eleven pixel wide asterisk
- stars brighter than 1.5 magnitude use a nine pixel wide asterisk
- stars brighter than 2.5 magnitude use a seven pixel wide asterisk

- stars brighter than 3.5 magnitude use a five pixel wide asterisk
- a plus sign is used for stars brighter than 4.5 magnitude
- a one pixel dot is used for stars fainter than magnitude 4.5

On the Mac, the stars are represented by filled circles with sizes correlated with the brightness of a given star. From brightest (largest) to dimmest (smallest):

- stars brighter than magnitude 0.0
- stars between magnitude 0.0 and magnitude 0.75
- stars between magnitude 0.75 and magnitude 1.5
- stars between magnitude 1.5 and magnitude 2.25
- stars between magnitude 2.25 and magnitude 3.0
- stars fainter than magnitude 3.0

The sky map plots all selected objects whose centers have a zenith distance of less than or equal to 90 degrees. This is different from the Rise/Set/Transit task which tabulates the times when the upper limbs of objects cross the horizon. In addition, the effects of refraction are not included in the sky map. This will have a negligible effect on the sky map (due to the map's scale) except for objects that are at or just below the theoretical horizon. For example, the Sun will not appear on a sky map which was generated for the time of sunrise as predicted by MICA's Rise/Set/Transit task. Instead, the Sun will appear on a sky map generated for a slightly later time.

While the Sky Map window is active, the following additional functions can be selected in the PC version by pressing the right mouse button over any object in the Sky Map or any name in the legend. As shown below, a dialog box will pop up with the following selectable functions

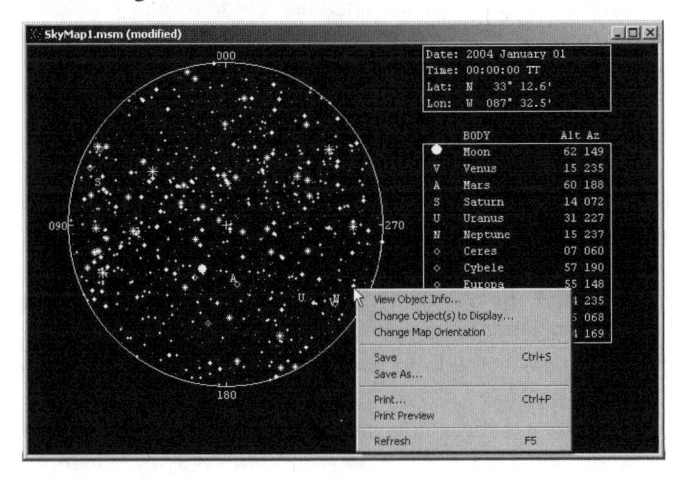

View Object Info (PC only) Displays the topocentric Right Ascension, Declination, Azimuth, Zenith Distance and Visual magnitude of the selected object. Note that the mouse cursor must be centered on an

object's symbol in order for this function to work.

Change Object(s) to Display Select a different group of objects for display.

Change Plot Orientation Reverses the Sky Map from East (Azimuth = 90°) on the left to East on the Right and vice versa.

Save Saves the Sky Map window to an opened Sky Map plot file. The MICA Sky Maps have a file extension of *.msm.

Save As Saves the Sky Map window to specified file name. The MICA Sky Maps have an file extension of *.msm.

Print Prints the Sky Map.

Print Preview Displays the Sky Map on the screen as it would appear printed.

Refresh Refreshes the Sky Map window.

The Sun's information is initially displayed on the lower right side of the Mac sky map. Other object information can be viewed by pressing the mouse cursor over any other symbol on the sky map. The entire list of objects displayed in the sky map can be viewed by pressing the 'Show List' button. The orientation of the sky map (East on left or right) can be reversed by pressing the 'East->' button.

3.4.6 Eclipses & Transits

This section calculates the local circumstances for solar and lunar eclipses, and for the transits of Mercury and Venus over a 250 year time span (1800 to 2050).

3.4.6.1 Solar Eclipse

The definition of an eclipse, given in *Explanatory Supplement to the Astronomical Almanac*, is when: "... the shadow of one body falls upon another and temporarily blocks out a portion of the solar illumination." In the case of a solar eclipse, it is the Moon's shadow that falls on the Earth. There are three types of solar eclipses; partial, total, and annular. Which is seen, if any, depends on the relative positions of the Sun, Moon, and Earth, but also on the observer's topocentric position.

To calculate the local circumstances of a solar eclipse, select Solar Eclipse(s) from the MICA Eclipses and Transits sub-menu. The Solar Eclipse dialog will look something like the figure below for the PC (left) and the Mac (right). The Mac Eclipse dialog box has two separate views for entry of the Eclipse date information, and Location information. Enter the year in which the eclipse occurs and then select the appropriate eclipse from the drop-down list. Next, enter your location (latitude and longitude).

Clicking OK on either of these panels creates a Solar Eclipse Output table as shown on the facing page.

The body of the table contains the time of each contact point, the Sun's topocentric position at that time, its Position and Vertex Angles. The time of sun rise or set is also noted in the table if it occurs during the eclipse. The summary at the bottom contains the Duration, Magnitude, and Obscuration.

The **Position Angle** of a given contact point on the solar limb is measured

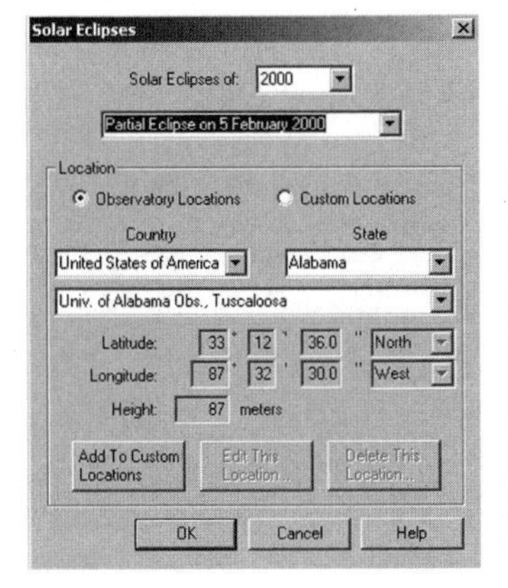

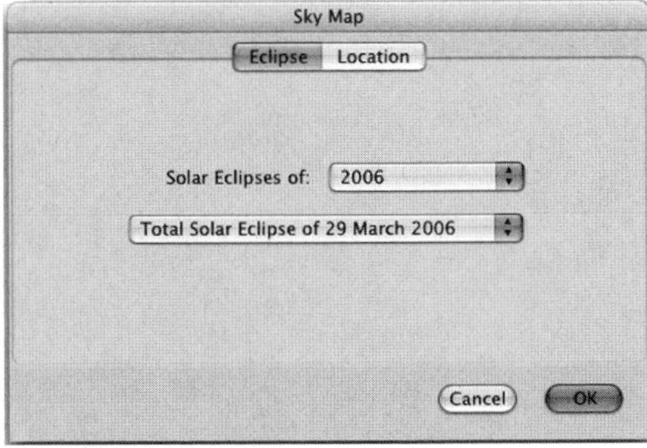

MICA Solar Eclipse Output Table

```
                 Solar Eclipse of 2006 March 29
             Sun in Partial Eclipse at this Location

                 National Obs. of Athens, Athens
        Location:  E 23°43'12.0", N37°58'24.0",    110m
          (Longitude referred to Greenwich meridian)

                                      Sun's          Position    Vertex
                     UT1        Altitude   Azimuth      Angle      Angle
                   h  m  s         °          °           °
Eclipse Begins    09:30:14.8     52.8      154.8       218.0      237.8
Maximum Eclipse   10:46:59.5     55.2      187.5
Eclipse Ends      12:03:28.3     49.4      217.5        57.0       28.4

          Duration:   2h 33m 13.5s
         Magnitude:   0.864
       Obscuration:  83.9%
```

eastward (counterclockwise) around the solar limb, from the point on the Sun that is farthest north.

Vertex Angle is similar to Position Angle, except that it is measured from the point on the Sun that has the highest local altitude.

Duration is the amount of time from the beginning of the eclipse to the end.

Duration of Annularity is the amount of time from the beginning of the central phase eclipse to the end of the central phase. For a total eclipse it will read Duration of Totality.

Magnitude of the eclipse is the fraction of the apparent diameter of the solar disk covered by the Moon at the time of greatest phase, expressed in units of solar diameter.

Obscuration is the percentage of the area of the apparent solar disk obscured by the Moon at maximum eclipse.

3.4.6.2 Lunar Eclipse

To calculate the local circumstances of a lunar eclipse, select Lunar Eclipse(s) from the MICA Eclipses and Transits sub-menu. The Lunar Eclipse dialog will look something like the figure below for the PC (left) and the Mac (right). The

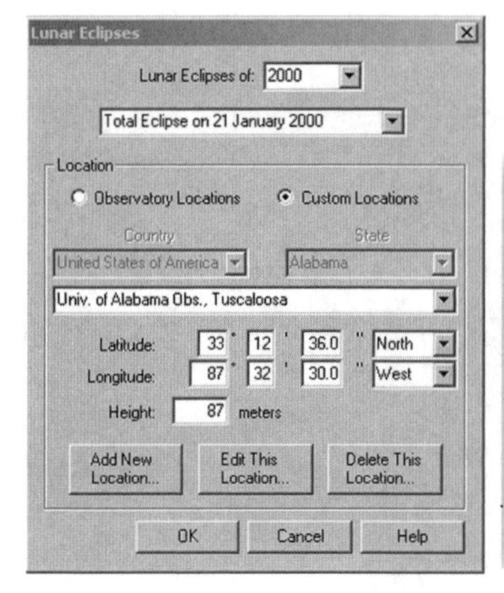

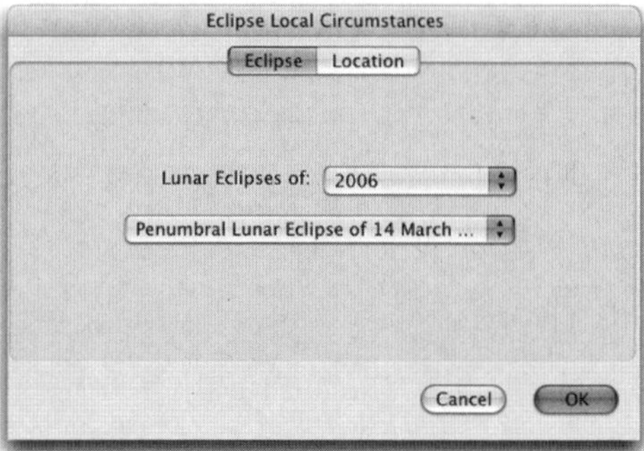

MICA Lunar Eclipse Output Table

```
Penumbral Eclipse of the Moon of 2006 March 14-15

                        Paris Obs., Paris
          Location:  E  2°20'12.0", N48°50'12.0",      67m
            (Longitude referred to Greenwich meridian)

                                             Moon's              Position
                              UT1       Altitude  Azimuth         Angle
                           d  h   m         °         °              °
Moon enters penumbra      14 21:21.6      33.2      130.7         158.8
Maximum Eclipse           14 23:47.5      43.5      175.1         209.0
Moon exits penumbra       15 02:13.5      35.5      221.0         259.3

          Penumbral Duration:    4h 51.9m
          Penumbral Magnitude:   1.057
```

Mac Eclipse dialog box has two separate views for entry of the Eclipse date information, and Location information. Enter the year in which the eclipse occurs and then select the appropriate eclipse from the drop-down list. Next, enter your location (latitude and longitude).

The body of the table contains a list of all lunar eclipse events visible from the selected locations. Where visible, the time of each contact point (Moon enters or exits the penumbra, Moon enters or exits the umbra, and/or the start/end of totality) and the time of maximum eclipse are listed. The time of moon rise or set is also noted in the table if it occurs during the eclipse. The Moon's topocentric position at that time, and the position angle of the Moon relative to the Sun are also included in the body of the table. The bottom of the table contains the duration time and magnitude of the eclipse.

3.4.6.3 Transit of Mercury

Technically, a transit is an eclipse. However with a transit, the body casting the shadow is not the Moon. Mercury and Venus are the only planets that can possibly cast a shadow on the Earth, The rules for whether or not a transit is seen, depends on the relative positions of the Sun, planet, and Earth, but also on the observer's topocentric position.

To calculate the local circumstances of a Mercury transit, select Transit(s) of Mercury from the MICA Eclipses and Transits sub-menu. Next, select the appropriate transit from the drop-down menu and the location. The Mac Transit of Mercury dialog box has two separate views for entry of the Transit date information, and Location information.

The body of the table contains the time of each contact point, the Sun's topocentric position, and two angles. The first, Position Angle, is similar to azimuth on the Earth, where zero degrees corresponds to due north from the Sun's center, ninety degrees is due east, and so forth. Angular Separation is the angular distance between the planet and the center of the Sun. the time of sun rise or set is also noted in the table if it occurs during the transit.

3.4.6.4 Transit of Venus

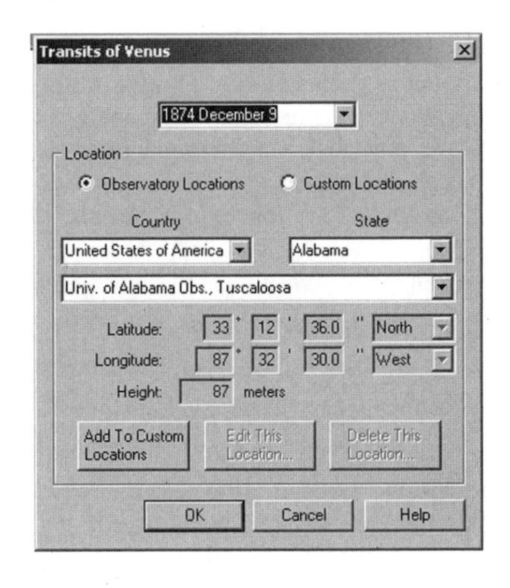

Technically, a transit is an eclipse. However with a transit, the body casting the shadow is not the Moon. Mercury and Venus are the only planets that can possibly cast a shadow on the Earth, The rules for whether or not a transit is seen depends on the relative positions of the Sun, planet, and Earth, but also on the observer's topocentric position.

To calculate the local circumstances of a Venus transit, select Transit(s) of Venus from the MICA Eclipses and Transits submenu. Next, select the appropriate transit from the drop-down menu and the location. The Mac Transit of Mercury dialog box has two separate views for entry of the Transit date information, and Location information. The PC dialog box is shown above.

The body of the table contains the time of each contact point, the Sun's topocentric position, and two angles. The first, Position Angle, is similar to azimuth on the Earth, where zero degrees corresponds to due north from the Sun's center, ninety degrees is due east, and so forth. Angular Separation is the angular distance between the planet and the center of the Sun. The time of sun rise or set is also noted in the table if it occurs during the transit, as it does in this case. Note that in the above tabulation the Transit Begin, Ingress Interior Contact, and the Least Angular Distance times are not listed because these events are not visible from this location.

MICA Venus Transit Output Table

```
                    Transit of Venus of 2012 June 06

                       U.S. Naval Obs., Washington
             Location:  W 77°04'00.0", N38°55'18.0",     92m
               (Longitude referred to Greenwich meridian)

                                                Sun's              Position   Angular
                                 UT1      Altitude   Azimuth       Angle    Separation
                               h  m  s        °          °           °           '
Transit Begins                22:03:57.5     25.9      279.1        41.2        16.2
Ingress Interior Contact      22:21:28.6     22.6      281.6        38.7        15.3
Sunset                        00:32          ----      300.7        -----       ----

             Duration:     2h 28m 28.3s
    Solar Semidiameter:   15' 45.7"
 Semidiameter of Venus:    0' 28.9"
```

3.4.7 Galilean Moons

This section generates three possible position types for Jupiter's Galilean satellites (Io, Europa, Ganymede, and Calisto). Position offsets of the moons from Jupiter are also tabulated. The information in this section utilizes the Galilean satellite ephemerides of Lieske (1977a). The MICA software is adapted from Lieske's original galsat50 fortran routine. See also Lieske (1977b).

On the PC, select Galilean Moons from the MICA Calculate menu. On the next dialog box, select one of following four options:

- Position Differences (in RA, Dec, Dist, and PA)
- Astrometric Geocentric Equator of J2000
- Apparent Geocentric Equator of Date
- Apparent Topocentric Local Horizon

On the next dialog box, enter the date and time, number of repetitions of the calculation, the Time System (TT or UT1), and the observer's location (for topocentric calculations only). The data Output format can be specified by the 'More Options' button.

On the Mac, select Galilean Moons from the MICA Compute menu. The Mac Galilean Satellites dialog box has five separate views for entry of the desired Position type calculation, Date/Time information, number of Repetitions, Location (for topocentric calculations only), and data Output format.

MICA tabulates the difference of the Apparent Geocentric Equator of Date satellite position minus Jupiter's position. The position differences are given in Right Ascension, Declination, angular separation, and position angle (PA). The position angle is defined so that 0° is due North with positive angles East of North and negative angles West of North. The Apparent Geocentric Equator of Date position of Jupiter is also listed above the satellite position differences.

Position Differences (in RA, Dec, Dist, and PA)

JOVIAN SYSTEM

Position Differences
Satellite - Jupiter

2004 Feb 02 00:00:00.0 (UT1)

	R.A.	Declination	Distance
	h m s	° ' "	A.U.
Jupiter	11 16 26.194	+ 6 08 54.35	4.575269337

	Delta RA	Delta Dec.	Separation	PA
	s	' "	' "	°
Io	+ 07.344	- 0 52.76	2 2.15	115.7
Europa	+ 07.124	- 0 44.50	1 55.76	112.7
Ganymede	- 19.583	+ 2 17.57	5 24.37	-64.8
Callisto	+ 31.203	- 3 44.70	8 39.18	115.8

Astrometric Geocentric Equator of J2000

JOVIAN SYSTEM

Astrometric Positions
Mean Equator and Equinox of J2000.0

2004 Feb 02 00:00:00.0 (UT1)

	Right Ascension	Declination	Distance
	h m s	° ' "	A.U.
Jupiter	11 16 13.106	+ 6 10 16.22	4.575269337
Io	11 16 20.452	+ 6 09 23.46	4.575865521
Europa	11 16 20.232	+ 6 09 31.73	4.571451917
Ganymede	11 15 53.520	+ 6 12 33.76	4.575497056
Callisto	11 16 44.314	+ 6 06 31.56	4.580452410

Apparent Geocentric Equator of Date

JOVIAN SYSTEM

Apparent Geocentric Positions
True Equator and Equinox of Date

2004 Feb 02 00:00:00.0 (UT1)

	Right Ascension	Declination	Distance
	h m s	° ' "	A.U.
Jupiter	11 16 26.194	+ 6 08 54.35	4.575269337
Io	11 16 33.538	+ 6 08 01.59	4.575865521
Europa	11 16 33.318	+ 6 08 09.85	4.571451917
Ganymede	11 16 06.610	+ 6 11 11.92	4.575497056
Callisto	11 16 57.396	+ 6 05 09.65	4.580452410

Apparent Topocentric Local Horizon

JOVIAN SYSTEM

Apparent Topocentric Positions
Local Zenith and True North

U.S. Naval Obs., Washington
Location: W 77°04'00.0", N38°55'18.0", 92m
(Longitude referred to Greenwich meridian)

2004 Feb 02 00:00:00.0 (UT1)

	Zenith Distance	Azimuth	Distance
	° ' "	° ' "	A.U.
Jupiter	36 06 17.2	209 07 18.5	4.575280250
Io	36 06 24.3	209 03 52.5	4.575876458
Europa	36 06 17.9	209 04 03.0	4.571462850
Ganymede	36 06 01.5	209 16 25.7	4.575507907
Callisto	36 06 48.5	208 52 43.2	4.580463422

SOLSTICES AND EQUINOXES

Year	Equinox Date		Time (UT)		Solstice Date		Time (UT)		Equinox Date		Time (UT)		Solstice Date		Time (UT)	
		d	h	m		d	h	m		d	h	m		d	h	m
2004	Mar	20	06	49	Jun	21	00	57	Sep	22	16	30	Dec	21	12	42
2005	Mar	20	12	33	Jun	21	06	46	Sep	22	22	23	Dec	21	18	35

3.4.8 Phenomena

Gives the dates of various astronomical phenomena (Solstices and Equinoxes, Moon phases, conjunctions, oppositions, greatest elongations of Mercury and Venus). A phenomena search feature is also available which generates a table similar to the 'Diary of Phenomena' tables contained in section A of *The Astronomical Almanac*.

3.4.8.1 Solstices/Equinoxes

MICA calculates the times of the Vernal and Autumnal Equinoxes and the Summer and Winter Solstices, for a range years. The equinox is defined to occur when the apparent geocentric longitude of the Sun is 0° (vernal equinox) or 180° (autumnal equinox). The equinox corresponds to two points on the celestial sphere at which the ecliptic intersects the celestial equator. The summer and winter solstices occur when the geocentric longitude of the Sun is 90° and 270°, respectively. To obtain Solstice/Equinox information select Phenomena from the MICA Calculate/ Compute menu. Next select Solstices/Equinoxes from the Phenomena submenu and then enter the desired date range.

3.4.8.2 Moon Phases

MICA will generate the times of the four 'primary' phases of the Moon for a range of dates. New Moon, first quarter, full Moon, and last quarter are defined as the times at which the excess of the apparent longitude of the Moon over that of the Sun is 0°, 90°, 180°, and 270°, respectively. The following describes how the Moon appears at each of the four phases:

PHASES OF THE MOON

Phase	Date (UT)	Time	Julian Date
		h m	d
Full	2004 Jan 07	15:40	2453012.1529
Last Quarter	2004 Jan 15	04:46	2453019.6983
New	2004 Jan 21	21:05	2453026.3784
First Quarter	2004 Jan 29	06:03	2453033.7522
Full	2004 Feb 06	08:47	2453041.8659
Last Quarter	2004 Feb 13	13:40	2453049.0691
New	2004 Feb 20	09:18	2453055.8873
First Quarter	2004 Feb 28	03:24	2453063.6417

- **New Moon** The Moon's hemisphere facing the Earth receives and reflects no direct sunlight.
- **First Quarter** One-half of the Moon appears to be illuminated by direct sunlight. The fraction of the Moon's disk that is illuminated is increasing.
- **Full Moon** The Moon's illuminated side is facing the Earth. The Moon appears to be completely illuminated by direct sunlight.
- **Last Quarter** One-half of the Moon appears to be illuminated by direct sunlight. The fraction of the Moon's disk that is illuminated is decreasing.

To obtain Moon Phase information, select Phenomena from the MICA Calculate/Compute menu. Next select Moon Phases from the Phenomena submenu and then enter the desired date range.

3.4.8.3 Conjunctions

MICA 2.0 tabulates several types of conjunctions for a range of dates. A solar conjunction occurs when the Sun and another celestial body (planet and /or asteroid) have the same geocentric longitude. A Lunar conjunction or a planetary conjuction occurs when the two celestial bodies have the same apparent geocentric right ascension. MICA tabulates the following conjunctions:

- The Sun with the major planets and/or selected asteroids.
- The Moon with the major planets, selected asteroids and/or selected bright stars.
- A major planet(s) with the other major planet(s), selected asteroids and selected bright stars.

The bright stars included in the MICA conjunction tabulations are Aldebaran, Antares, Pollux, Regulus, and Spica. The times and dates of the conjunctions are tabulated, in addition to the apparent geocentric right ascension and the separation, in degrees.

```
                       CONJUNCTIONS WITH SUN
                 From: 2004 Jan 01   To: 2004 Jun 29

Planet                     Date             Right Ascension     Separation
                           (UT)
                              d    h  m        h  m  s              °
Neptune            2004 Feb  2   9:27       21 01 23          0.03 South
Uranus             2004 Feb 22   2:07       22 20 01          0.67 South
Mercury superior   2004 Mar  4   1:43       23 03 05          1.61 South
Mercury inferior   2004 Apr 17   1:05        1 38 51          1.74 North
Venus   inferior   2004 Jun  8   8:43        5 07 25          0.18 South
Mercury superior   2004 Jun 18  21:24        5 50 59          1.04 North
```

To obtain Conjunction information, select Phenomena from the MICA Calculate/Compute menu and then select Conjunctions from the Phenomena submenu. There are three Conjunctions submenu choices: Sun, Moon, and Planets/Other. On the next dialog box enter the desired date range and the object(s) to search for conjunctions with.

3.4.8.4 Oppositions

OPPOSITIONS WITH SUN
From: 2004 Jan 01 To: 2004 Feb 29

Object	Date (UT)		RA	Declination
	d	h m	h m s	° ′ ″
Ceres	2004 Jan 09	13:53	7 25 49	+30 07 54
Hygiea	2004 Jan 07	14:48	7 12 00	+22 07 46
Hebe	2004 Jan 12	23:06	7 27 53	+10 10 34
Eunomia	2004 Feb 11	11:47	9 26 20	+ 5 46 08

MICA generates oppositions of the Sun with the major planets and/or selected asteroids for a range of dates. An opposition is defined to occur when the apparent geocentric Right Ascension of the Sun differs by 180° from the celestial object. For each object, the times and dates of oppositions are tabulated, in addition to the apparent geocentric right ascension and declination.

To obtain Opposition information, select Phenomena from the MICA Calculate/Compute menu and then select Oppositions from the Phenomena submenu. On the next dialog box enter the desired date range and the object(s) to search for oppositions with.

3.4.8.5 Greatest Elongations

GREATEST ELONGATIONS
From: 2004 Mar 01 To: 2004 Aug 29

Planet	Date (UT)		Elongation		Mag.
	d	h m	°		
Mercury	2004 Mar 29	12 27	18.9	East	-0.1
Venus	2004 Mar 29	16 40	46.0	East	-4.3
Mercury	2004 May 14	20 36	26.0	West	0.4
Mercury	2004 Jul 27	03 29	27.1	East	0.4
Venus	2004 Aug 17	18 32	45.8	West	-4.2

MICA will tabulate the times and dates of greatest elongations of Mercury and/or Venus from the Sun. The elongation is the geocentric angle between the Sun and Mercury or Venus. Greatest Elongation occurs when this angle is at a maximum. The Elongation angle (in degrees), the direction (East or West) and the approximate magnitude of Mercury or Venus are also tabulated.

To obtain Greatest Elongation information, select Phenomena from the MICA Calculate/Compute menu and then select Greatest Elongations from the Phenomena submenu. On the next dialog box enter the desired date range and object (Mercury and/or Venus).

3.4.8.6 Phenomena Search

MICA tabulates a listing of geocentric phenomena, similar to the 'Diary of Phenomena' contained in section A of *The Astronomical Almanac*. The MICA Diary tabulates the dates and times of the equinoxes and solstices; primary phases of the

Moon; geocentric conjunctions of the Sun, Moon, and/or planets with the major planets, selected asteroids, and selected bright stars (Aldebaran, Antares, Pollux, Regulus, and Spica); oppositions of the Sun with the major planets and selected asteroids; and greatest elongations of Mercury and Venus. The dates and times are tabulated as well as the object separation, where relevant. Note that the separation, if applicable, is given relative to the latter object. For example, on 15 March 2004, Juno is 16.68° north of the Moon.

```
                          PHENOMENA SEARCH
                 From: 2004 Mar 14  To: 2004 Mar 27

Phenomena found:                              Date              Comment
                                                    d  h
Juno in conjunction with Moon             2004 Mar 15 07    16.68° North
Psyche in conjunction with Moon           2004 Mar 16 07     7.13° North
Neptune in conjunction with Moon          2004 Mar 17 09     5.33° North
Metis in conjunction with Moon            2004 Mar 17 20     1.54° North
Vesta in conjunction with Moon            2004 Mar 17 12     3.91° North
Uranus in conjunction with Moon           2004 Mar 18 20     4.39° North
Equinox                                   2004 Mar 20 06
New Moon                                  2004 Mar 20 23
Flora in conjunction with Moon            2004 Mar 20 17     0.08° North
Mercury in conjunction with Moon          2004 Mar 22 04     3.60° North
Cybele in conjunction with Moon           2004 Mar 23 01     1.24° South
Venus in conjunction with Moon            2004 Mar 24 21     2.16° North
Mars in conjunction with Moon             2004 Mar 25 23     0.81° South
Europa in conjunction with Moon           2004 Mar 25 02     6.74° South
Pallas in conjunction with Moon           2004 Mar 25 00    27.75° South
Interamnia at opposition                  2004 Mar 25 04
Aldebaran in conjunction with Moon        2004 Mar 26 15     8.11° South
Pallas in conjunction with Venus          2004 Mar 27 10    29.86° South
Venus in conjunction with Pallas          2004 Mar 27 10    29.86° North
```

To search for phenomena information on the PC, select Phenomena from the MICA Calculate menu and then select Search from the Phenomena submenu. On the next dialog box, enter the desired date range. This search will include all Solstice/Equinox, Moon Phase, Conjunction, Opposition, and Greatest Elongation events/dates that occurred during the date range.

To search for phenomena information on the Mac, select Phenomena from the MICA Compute menu and select Search from the Phenomena submenu. The Mac dialog box consists of six views: Interval — for specification of the date range; Sun — for solar conjunction object selection; Moon — for lunar conjunction object selection; Conjunctions — for planetary conjunction object selection; and Other — for solstice/equinox, moon phase and greatest elongation computations. Use the 'Search All' button to find all possible phenomena. Use the 'Clear All' button to clear the search criteria.

3.5 Preferences Menu Commands (PC only)

Several of the MICA input and output fields may be preset by use of the Set Preferences utility. These settings can be saved for later use with any particular calculate menu function.

3.5.1 System Time at Startup

When this field is checked, the Computer System Time is used at startup of MICA for the initial calendar date and time fields. When this field is not checked (the default mode), the calendar date and time is saved upon EXIT from MICA and is used for the initial calendar date and time fields the next time MICA is started.

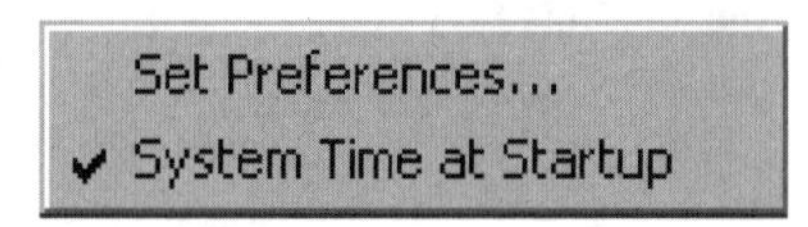

3.5.2 Set Preferences (PC only)

Several of the MICA input and output fields may be preset by use of the Set Preferences utility. These settings will be saved for use with a particular calculate menu function.

Date System Two Date Systems can be selected: Julian Date or Calendar Date (e.g., day/month/year and hour:minute:second). These settings are used for the date and time input fields and for the output tables.

Time System Two Time Systems can be selected: TT or UT1. These settings are used for the time scale input field and for the output tables.

Twilights Three twilight types can be selected: Civil, Nautical, or Astronomical for the Rise/Set/Transit function. The beginning and ending twilight times are included in the Rise/Set/Transit output tables for the sun.

Output Format Object positions are displayed in Decimal Format or in Degrees, Minutes, Seconds and/or Hours, Minutes, Seconds (DMS/HMS) format. This applies to the Positions, Configurations, and Galilean Moons calculate menu functions.

Location Manager This dialog box sets the observer's location (latitude, longitude and height) on the earth. Choose either to use the MICA internal database of Astronomical Observatories (based on the one contained in section J of *The Astronomical Almanac*) or enter a custom location. The custom locations may be saved, edited, and/or deleted via the Add New Location, Edit This Location, and/or Delete This Location boxes.

Time Zone Select the appropriate Time Zone or enter the time zone offset from Greenwich. Note that a fractional time zone offset maybe used for certain locations (e.g., 9.5 hours for central Australia). Also select whether Daylight Saving time is used or not. This field is only used for Rise/Set/Transit calculations. Standard time (or Zone Time) is given by the expression: Standard Time = Universal Time + Time Zone Offset.

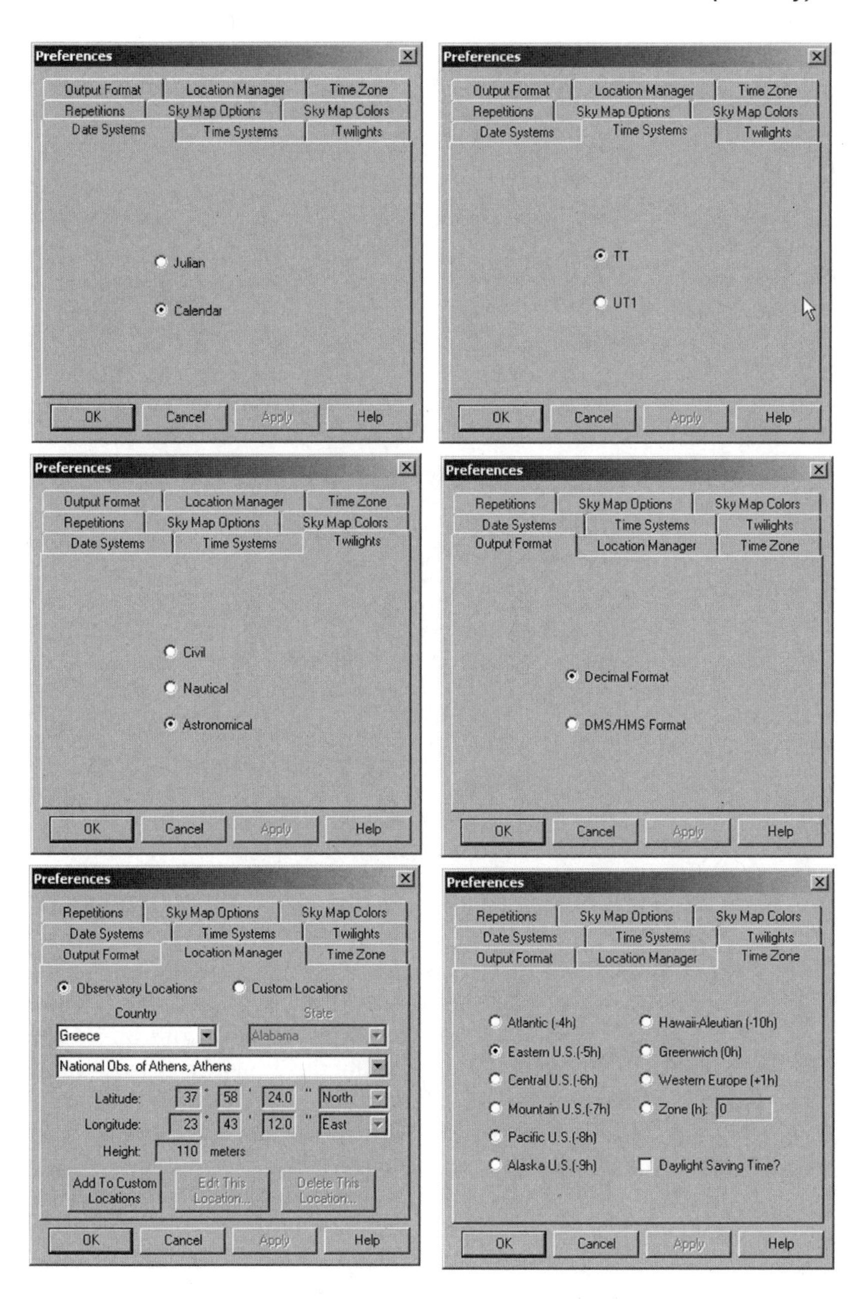

Repetitions Select the number of repetitions for a given calculation and the length of time between each repeated calculation.

Sky Map Options Select the Objects to Display (Planets, Asteroids, and/or Stars). If Stars are to be displayed, then either select 'All Stars' or specify a magnitude range. The plot orientation may also be adjusted

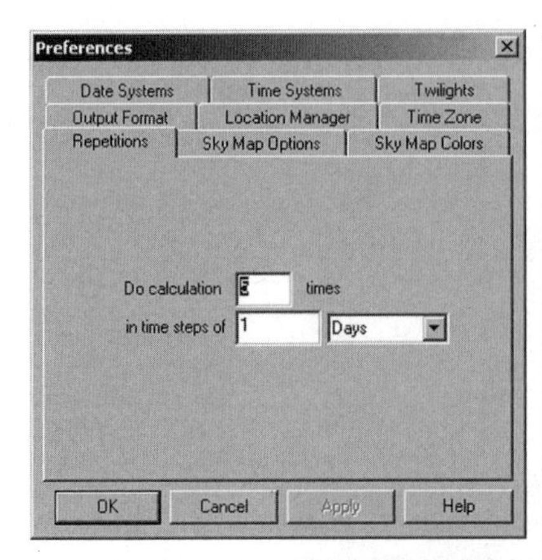

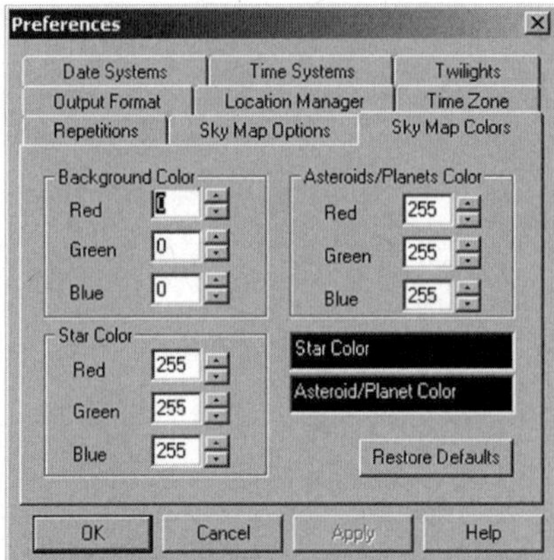

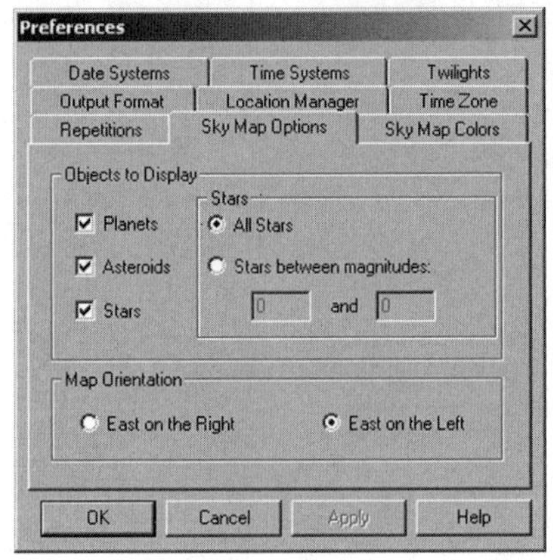

so that East (90° Azimuth) is either on the right or the left side of the plot.

Sky Map Colors for the background sky, stars, and asteroids/planets are selected by choosing the intensity of red, green, and blue for each of the three object types. The results are displayed in windows on the lower right of the dialog box. The values for these numbers range from zero to 255. There are NO other restrictions placed on the colors one can choose. This means that one can choose to plot black stars on a black background, and MICA will let one do that. If one wishes to reset the colors to their factory default, one can click the 'Restore Defaults' button

3.6 Settings (Mac only)

Date System Two Date Systems can be selected: Julian Date or Calendar Date (e.g. day/month/year and hour:minute:second). These settings are used for the date and time input fields and for the output tables.

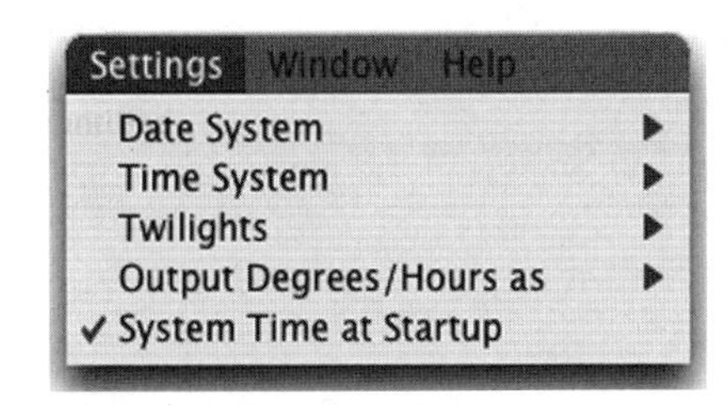

Time System Two Time Systems can be selected: TT or UT1. These settings are used for the time scale input field and for the output tables.

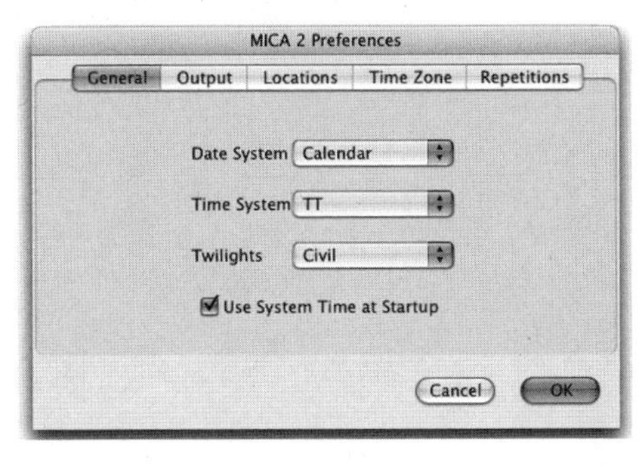

Twilights Three twilight types can be selected: Civil, Nautical, or Astronomical for the Rise/Set/Transit function. The beginning and ending twilight times are included in the Rise/Set/Transit output tables.

Output Degrees/Hours as Object positions are displayed in Decimal Format or in Degrees, Minutes, Seconds and/or Hours, Minutes, Seconds (DMS/HMS) Format. This applies to the Positions, Configurations, and Galilean Moons functions.

Preferences Allows the user to set certain values that are used the Compute functions. **Under OS X, this command appears under the MICA menu.**

The Preferences pane can be used to set the values seen under the Settings menu; select the output format of the data; select a location (or enter a new location into the Custom Locations database); enter the time zone (Used only in Rise/Set and the Sky Map functions) and the number of repetitions to perform

3.7 Window Menu Commands (PC only)

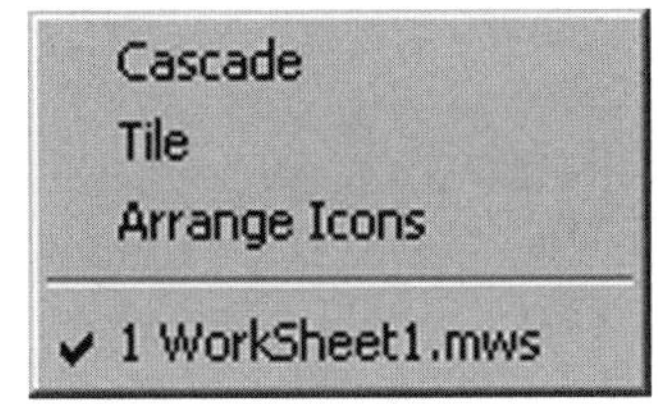

The Window menu offers the following commands, which enables one to arrange multiple views of multiple documents in the application window:

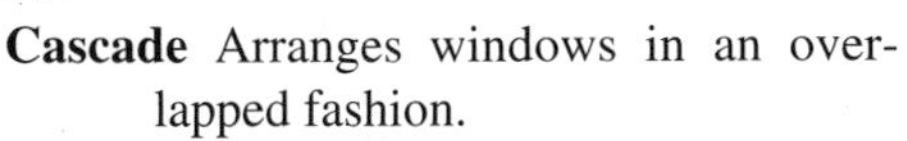

Cascade Arranges windows in an overlapped fashion.

Tile Arranges windows in non-overlapped tiles.

Arrange Icons Arranges icons of minimized windows.

Worksheet1.mws, ... or Skymap1.msp, ... Goes to specified Worksheet or Skymap window.

Chapter 4

MICA Dialog Boxes for the PC

4.1 Asteroid Dialog Box

The Asteroid Dialog Box lists the 15 asteroids for which MICA 2.0 has ephemerides. See Sections 3.4.1 (Positions) on page 19 and 3.4.3 (Rise/Set/Transit) on page 27.

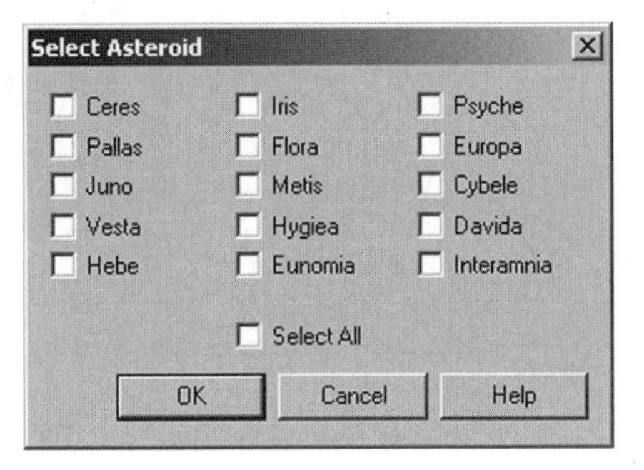

4.2 Bright Star Dialog Box

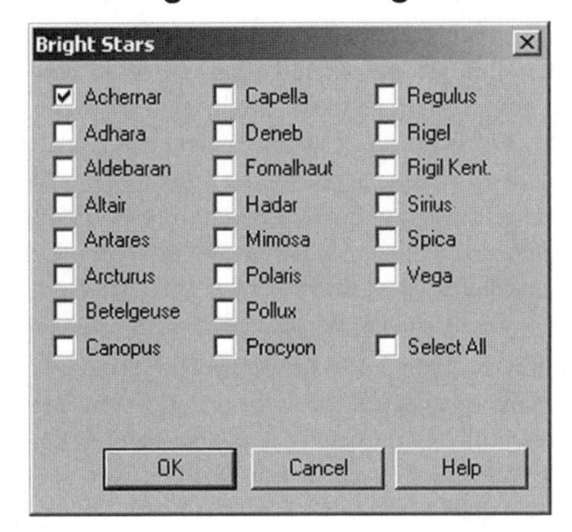

The Bright Star Dialog Box lists the 22 bright stars that can be used in MICA 2.0. This is the same list that was used in MICA 1.5. See Sections 3.4.1 (Positions) on page 19 and 3.4.3 (Rise/Set/Transit) on page 27.

4.3 Catalog Object Search

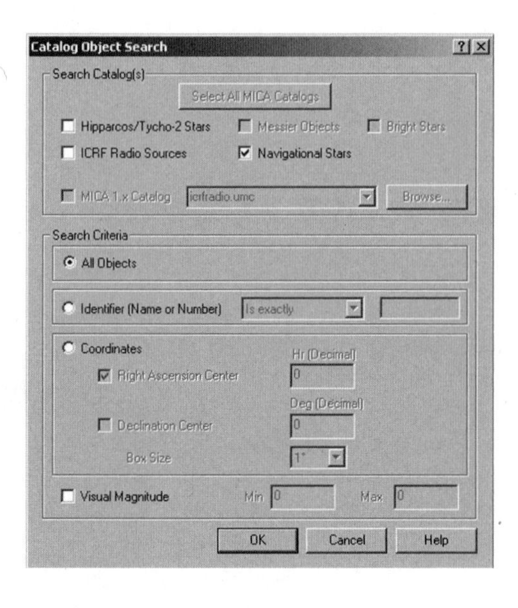

Search Catalog(s): Select the catalog(s) which are to be searched. Multiple catalogs may be selected. Several external catalogs have been included with the MICA 2.0 release. See Sections 3.4.1 (Positions) on page 19 and 3.4.3 (Rise/Set/ Transit) on page 27.

Search Criteria: MICA supports the following types of catalog searches:

- **All Objects:** This will search for ***all*** objects in the catalogs selected. It is strongly recommended that this be used in conjunction with the Visual Magnitude Range search as described below so as to limit the number of objects found.
- **Identifier (Name or Number):** This will search for a specific catalog identifier. Additional object name syntax information is available for the *MASC catalog*, the *ICRF Radio Source catalog*, the *Navigational Star catalog*, and the *Bright Stars catalog*. Two types of searches can be performed:
    - **Is exactly:** This search will match the entered object name or number exactly. MICA will properly format the input object name before searching a given catalog (with the exception of imported MICA 1.x catalogs, where the search is white-space sensitive but case insensitive). Extra blank white spaces are automatically removed or added, as appropriate, before performing the search. MICA object name searches are also case insensitive (i.e., the following searches are equivalent: 'Alf Ori,' 'ALF ORI6,' and/or 'alf ori').
    - **Starts with:** This will match all objects with the specified initial characters and with the same spacing format as the catalog identifiers (i.e., white spaces are not removed or added before the search is performed). This search is also case insensitive. For example, 'alf' will return all stars with names that begin with 'alf' such as, 'alf Ori,' or 'alf CMa'. Another example, 'HIP 2' will find nothing because the Hipparcos ID format is 'HIPbNNNNNN,' and there are no Hipparcos ID's in the *MICA* catalog with ID's greater than 'HIP 200000,'

where 'b' is a blank and 'NNNNNN' is the 6 digit Hipparcos ID number.

- **Coordinates:** Specify the Right Ascension (in hours, minutes, seconds) and Declination (in degrees, arcminutes, arcseconds) of the center of the region (box) to be searched. Box sizes range from 15" (arcminutes) to 15 degrees. Strips centered around a specific Right Ascension or Declination may also be searched.
- **Visual Magnitude Range:** Specify the range of magnitudes to be searched. This field can be used to limit the number of objects returned from the above 'All Object', 'Identifier' or 'Coordinates' searches. For example, to obtain a list of the brightest stars in the MASC catalog, select 'All Objects' and specify a Visual Magnitude Range of −1.5 to 2.0.

Warning: MICA does not limit the total number of catalog entries that may be returned for a specific query. Thus, it is possible to search and attempt to retrieve the *entire* MASC catalog (over 200,000 entries). This may take an extremely long time! It is strongly recommended that small query sizes be used.

4.4 Date Systems

Dates and times of astronomical events can be displayed in calendar or Julian date form. Calendar is the familiar year/month/day/hour/minute/second format. Julian date is a single double precision number that varies smoothly with time.

4.5 Date/Time/Location

Input data is arranged in four groups: Date and Time, Repetitions, Time Systems, and Location Manager.

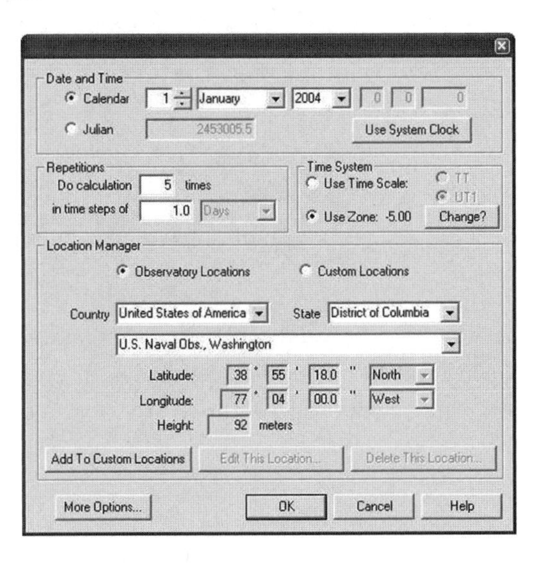

Date and Time gives one the option of using Julian or calendar date. The user has the option of using the computer's system clock to set the time.

Repetitions is straightforward, except in the case of rise/set calculations. Note that the time step can be a non integer number, but rise/set insists that the time steps be in increments of days, one day being the smallest increment. Whatever one enters for time step will be rounded to integral days for rise/set calculations (and only for those calculations; the actual value one entered will be saved).

Time Systems (see also the Time Systems tab of the preferences dialog box) is selected here. TT and UT1 are defined elsewhere. Note that the "Use Zone Time" option implies the use of UTC, which is not used by MICA 2.0. Thus, only rise/set and sky map calculations allow the use of zone time, as their output is accurate to the nearest minute at best, and ΔUT1 is always less than 0.9 seconds in absolute value.

4.6 Galilean Moons

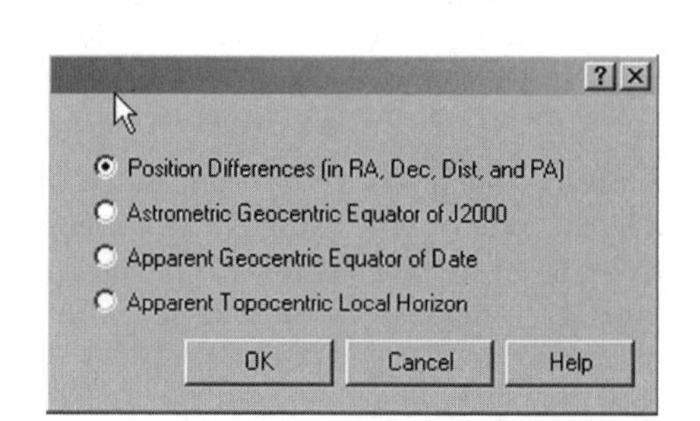

There are four calculation options to choose from:

- Position Differences (in RA, Dec, Dist, and PA)
- Astrometric Geocentric Equator of J2000.0
- Apparent Geocentric Equator of Date
- Topocentric Horizon

See Section 3.4.7 (Galilean Moons) on page 46.

4.7 Greatest Elongation

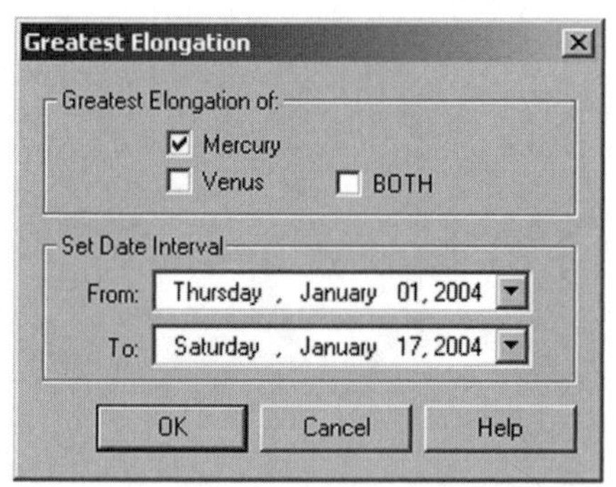

To do a Greatest Elongation calculation, select the desired objects (Mercury, Venus or Both) and date interval. See Section 3.4.8.5 (Greatest Elongations) on page 50.

4.8 Location Manager

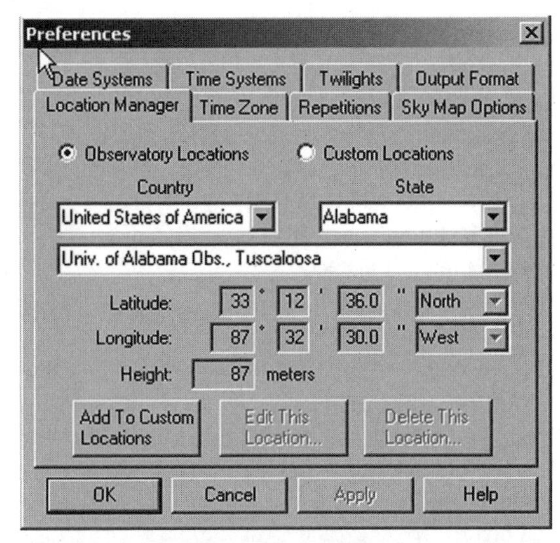

The location manager is used in any of the MICA computations involving topocentric calculations, including the Sky Map utility. The location manager consists of a pair of databases, one provided by the USNO, Observatory Locations, and one the user constructs, Custom Locations.

The former is a list of 469 observatories worldwide. This is the same list that appears in *The Astronomical Almanac*. To select one of the observatories, click the 'Observatory Locations' radio button, and then select a country, a state, and an observatory from the drop-down lists. If you wish to add a particular observatory location to the Custom Locations list, simply click the 'Add to Custom Locations' button.

The observatory database is read-only, so the 'Edit This Location' and the 'Delete This Location' buttons are greyed out.

To create a custom location click on the 'Custom Locations' radio button, then click on the 'Add New Location' button.

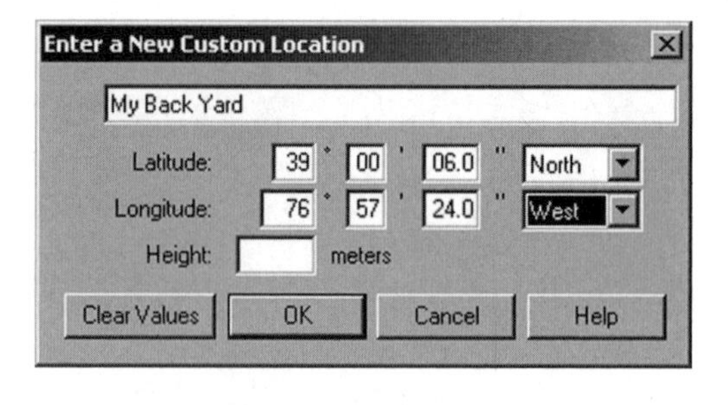

If you wish to enter a throw away location that will be used only once, enter it here. If you wish to enter a location that will be used more than once, click on the 'Add New Location' button. Note that in order to be able to recall this location from a drop down list later, it is necessary to give this location a title. It can be anything you want.

To Enter a New Custom Location, type the latitude, longitude, and height (above the geoid), and give the location a unique name. When one presses the 'OK' button, this location will be displayed in the Location Manager dialog box.

To Edit a Custom Location make whatever changes deemed necessary; then press the 'OK' button. This location will be displayed in the Location Manager dialog box.

To Delete a Custom Location, press the 'Delete This Location' button.

4.9 Lunar Conjunctions

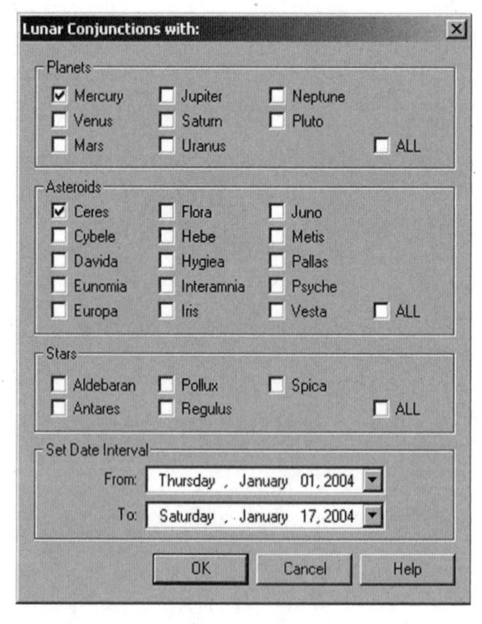

Lunar conjunctions are between the Earth, Moon, and a third, conjuncting body. This dialog box allows one to choose amongst a selection of all possible planets (the Earth excluded, of course), all possible asteroids (the one's for which we have ephemerides), and a small selection of stars. The date interval must also be specified. See Section 3.4.8.3 (Conjunctions) on page 49.

4.10 Lunar Eclipses

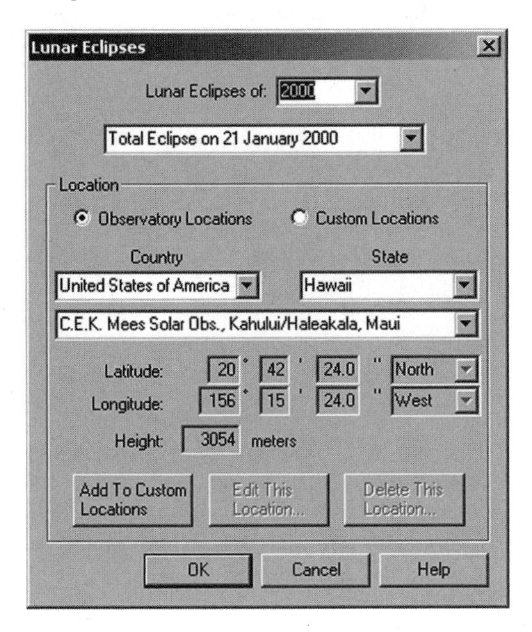

To calculate the local circumstances of a lunar eclipse, select Lunar Eclipses from the MICA Eclipses and Transits sub-menu. Enter the year in which the eclipse occurs, then select the appropriate eclipse from the drop-down menu. Next, enter your location (latitude and longitude). See Section 3.4.6.2 (Lunar Eclipse) on page 43.

4.11 Moon Phases

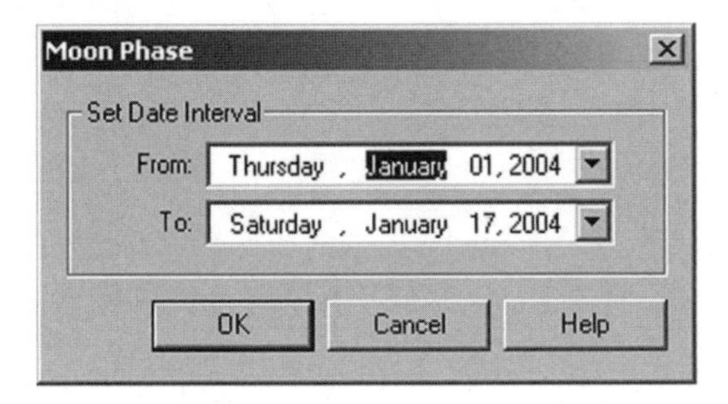

Pick an end search date, and MICA will display all Moon phases for that time interval. Note that they are independent of observer's location. See Section 3.4.8.2 (Moon Phases) on page 48.

4.12 Output Format

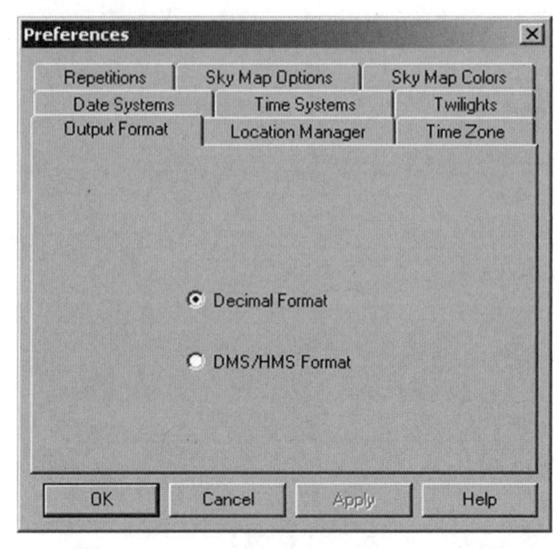

You can specify the numerical output for most types of calculations in decimal format, or in DMS (degree, arc minutes, arc seconds) / HMS (hours, time minutes, time seconds).

4.13 Oppositions

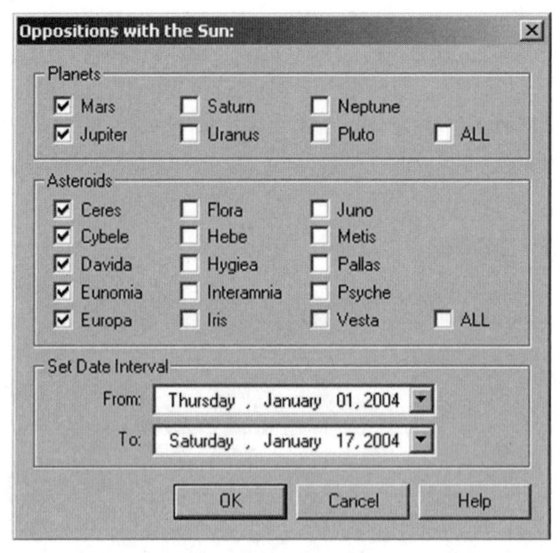

Oppositions between the Sun and other solar system objects can be calculated using this dialog box. All planets or asteroids can be selected (or deselected) by using the "ALL" button. See Section 3.4.8.4 (Oppositions) on page 50.

MICA generates oppositions of the Sun with the major planets and/or selected asteroids for a range of dates. An opposition is defined to occur when the apparent geocentric longitude of the Sun differs by 180° from the celestial object. For each object, the times and dates of oppositions are tabulated, in addition to the apparent geocentric right ascension and declination.

4.14 Phenomena Search

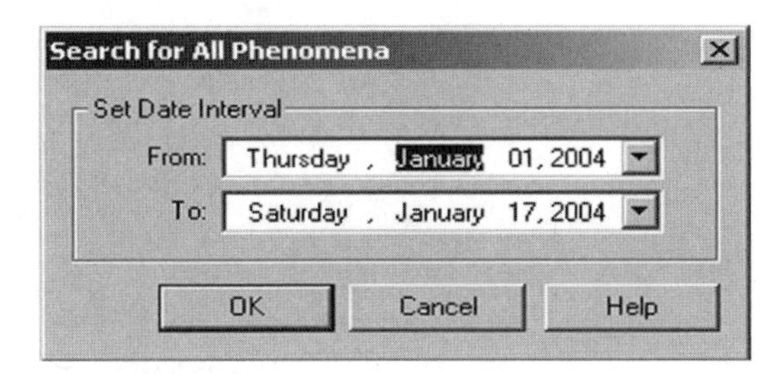

Enter begin and end dates to be able to search all phenomena during that time interval. See Section 3.4.8 on page 48 for additional phenomena information. See Section 3.4.8.6 (Phenomena Search) on page 50.

4.15 Physical Ephemeris

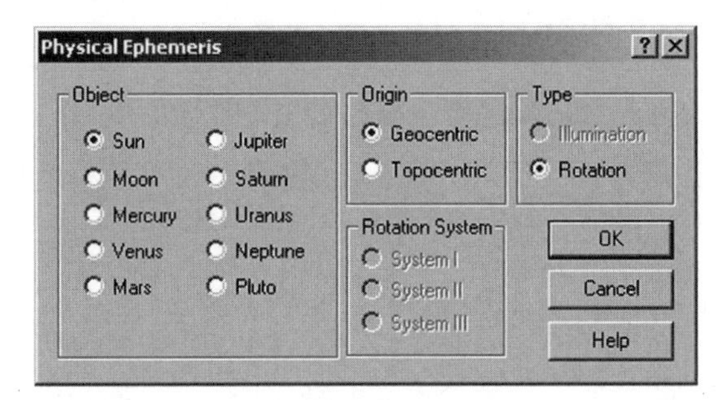

To do a physical ephemeris calculation, select an object, an origin, a rotation system (possible only when doing a Jupiter rotation), and a type (illumination or rotation). See Section 3.4.4 (Physical Ephemeris) on page 31.

4.16 Planetary Conjunctions

This dialog box calculates conjunctions of a major planet (in this case, Mercury) with the given list of planets, asteroids, and stars. The list of asteroids consists of all those for which MICA 2.0 has ephemerides. See Section 3.4.8.3 (Conjunctions) on page 49.

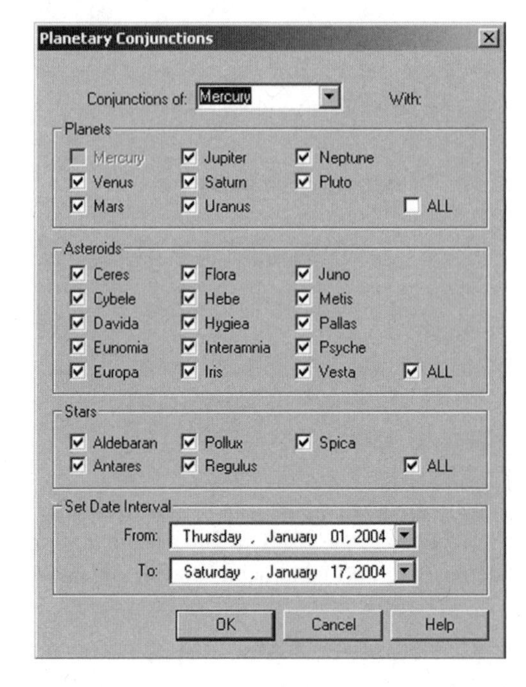

4.17 Positions Dialog Box

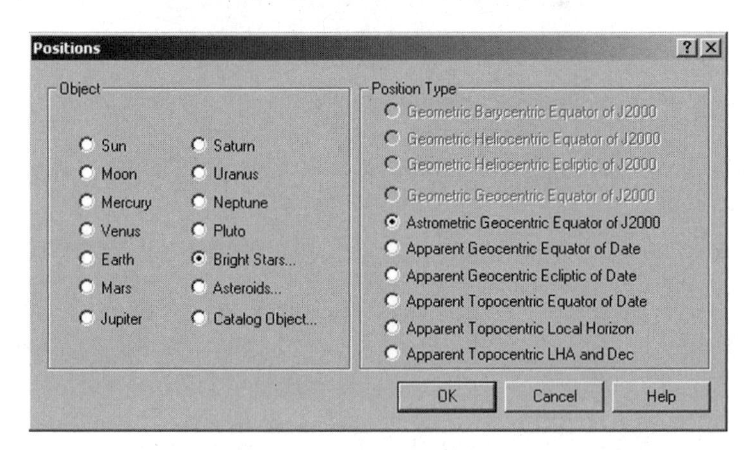

This dialog box is pretty straightforward. The only two things to note are that the position types allowed depends on which object is selected, and that more information on what the various position types mean can be found in chapter on MICA Position Calculations. See Section 3.4.1 (Positions) on page 19.

4.18 Repetitions

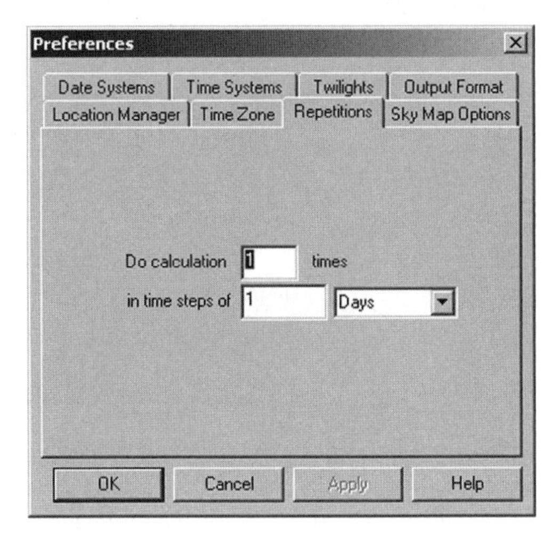

For many types of calculations, it is necessary to specify how many times you wish to perform the calculation, and over what time interval. This is where selections are made. Please note that, when doing rise/set calculations, the increment will be rounded to the nearest day (rounding up to one day if the interval was less than that).

4.19 Sky Map Colors

This property page gives the user some control over the colors displayed on the sky map. See Section 3.4.5.3 (Sky Map Configurations) on page 39.

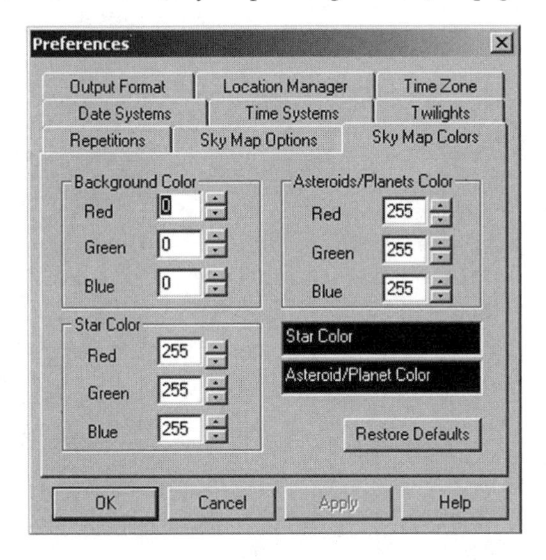

Colors for the background sky, stars, and asteroids/planets are selected by choosing the intensity of red, green, and blue for each of the three object types. The results are displayed in windows on the lower right of the dialog box. The values for these numbers range from zero to 255. There are NO other restrictions placed on the colors one can choose. This means that one can choose to plot black stars on a black background, and MICA will let one do that. If one wishes to reset the colors to their factory default, one can click the 'Restore Defaults' button.

4.20 Sky Map Object Window

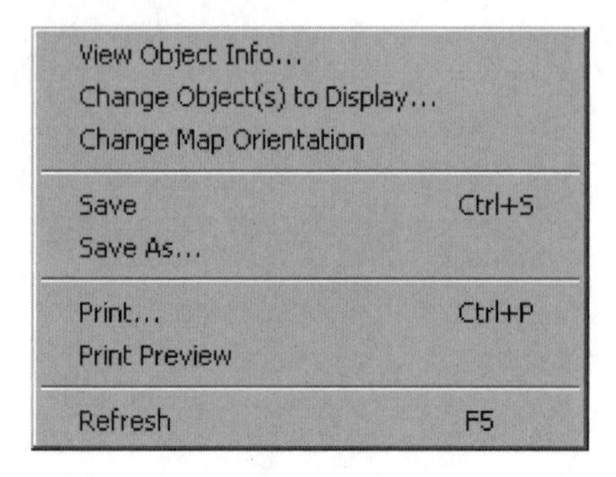

Once Sky Map is displayed right clicking on a object brings up the following dialog box. See Section 3.4.5.3 (Sky Map Configurations) on page 39.

This window displays the data for each star (if any) on the Sky Map if you right click on the symbol and select Object Info. The data displayed is the common name, the right ascension, the declination, the azimuth, the zenith distance, and the magnitude of the stellar object

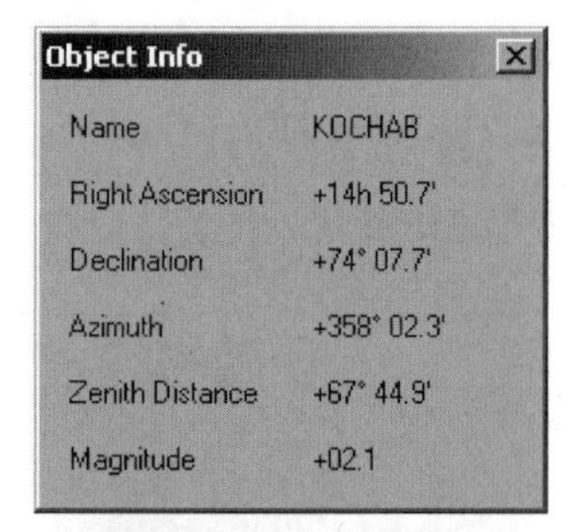

This window displays the data for the Moon, if it appears on the sky map and you right click on the symbol and select Object Info. The data is similar to that for stars, except for the magnitude, which include the percent illumination and the addition of semidiameter.

Object Info

Name	Moon
Right Ascension	+10h 48.8'
Declination	+09° 51.2'
Azimuth	+102° 15.4'
Zenith Distance	+59° 15.9'
Magnitude	-11.1 (87.2% full)
Semidiameter (")	890.4

This window displays the data for the Sun, planets, and asteroids, if they appear on the sky map and if you right click on the symbol and select Object Info. The data is similar to that for the Moon, except for the magnitude, which does not include the percent illumination.

Object Info

Name	Saturn
Right Ascension	+08h 45.1'
Declination	+18° 44.3'
Azimuth	+121° 05.3'
Zenith Distance	+30° 46.5'
Magnitude	-0.2
Semidiameter (")	10.2

4.21 Sky Map Options

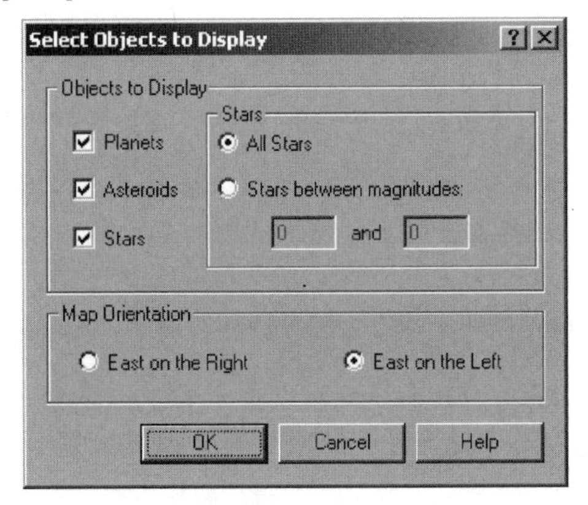

Input data is arranged in two groups: Objects to Display and Map Orientation.

Under Objects to Display, the Sun and Moon will plot automatically, if they are visible, but one can choose to not plot the planets, asteroids, and stars. If one wishes to plot stars in a range of magnitudes, one has that option as well.

The Map Orientation option allows for users who prefer to have their plots display properly when held up to the sky (East on the Left), or when held facing the ground (East on the Right). See Section 3.4.5.3 (Sky Map Configurations) on page 39.

4.22 Solar Conjunctions

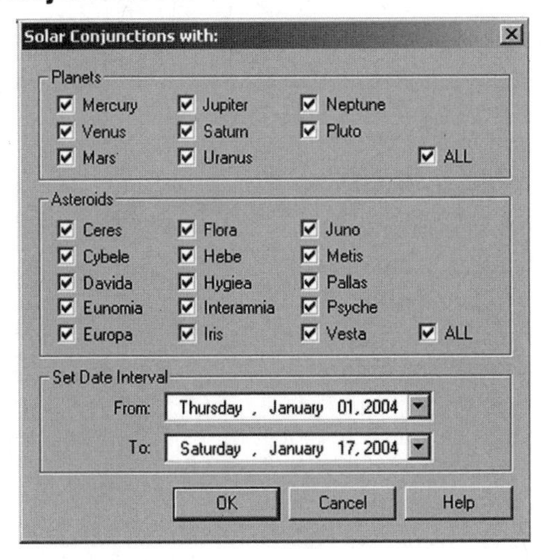

Conjunctions between the Sun and other solar system objects can be calculated using this dialog box. One can select (or deselect) all planets or asteroids by using the "ALL" button or a subset by clicking on individual objects. See Section 3.4.8.3 (Conjunctions) on page 49.

4.23 Solar Eclipse

To calculate the local circumstances of a solar eclipse, select Solar Eclipses from the MICA Eclipses and Transits sub-menu. Enter the year in which the eclipse occurs, then select the appropriate eclipse from the drop-down menu. Next, enter your location (latitude and longitude). The Position Angle of a given contact point on the solar limb is measured eastward (counterclockwise) around the solar limb, from the point on the Sun that is farthest north. See Section 3.4.6.1 (Solar Eclipse) on page 42.

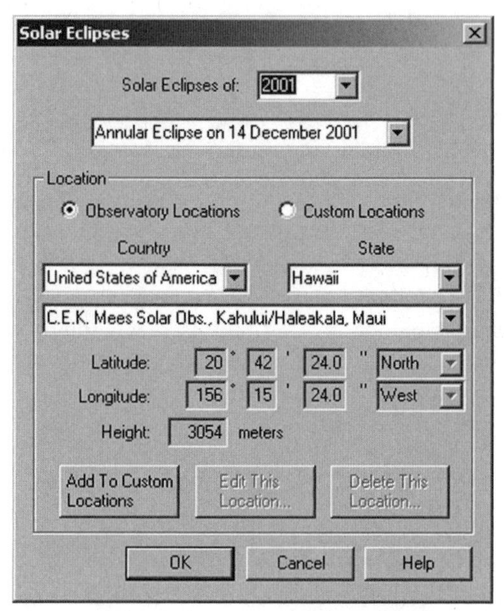

4.24 Solstices/Equinoxes

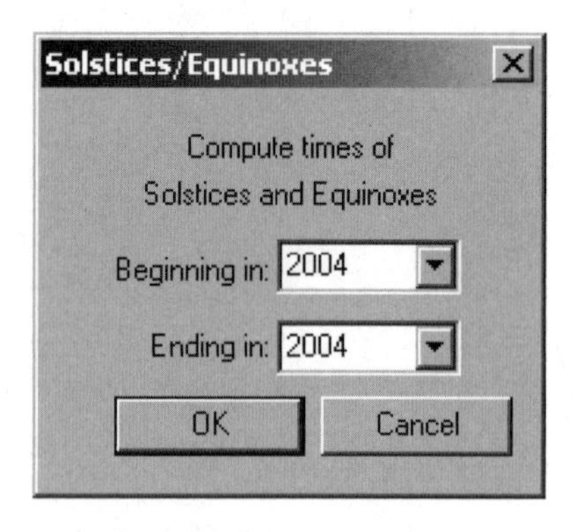

MICA 2.0 calculates the times of the Vernal and Autumnal Equinoxes and the Summer and Winter Solstices, for a range years. The equinox is defined to occur when the apparent geocentric ecliptic longitude of the Sun is 0° (vernal equinox) or 180° (autumnal equinox). The equinox corresponds to two points on the celestial sphere at which the ecliptic intersects the celestial equator. The summer and winter solstices occur when the geocentric ecliptic longitude of the Sun is 90° and 270°, respectively. See Section 3.4.8.1 (Solstices/Equinoxes) on page 48.

4.25 Transit of Mercury

To calculate the local circumstances of a Mercury transit, select Transits of Mercury from the MICA Eclipses and Transits sub-menu. Enter the year in which the eclipse occurs, then select the appropriate eclipse from the drop-down menu. Next, enter your location (latitude and longitude). See Section 3.4.6.3 (Transit of Mercury) on page 44.

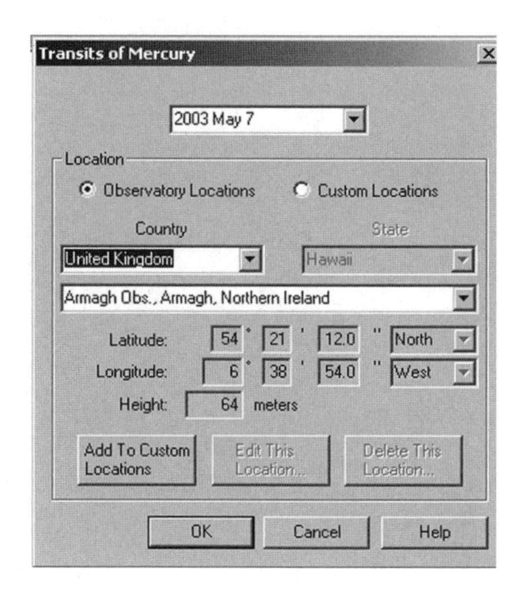

4.26 Transit of Venus

To calculate the local circumstances of a Venus transit, select Transits of Venus from the MICA Eclipses and Transits sub-menu. Enter the year in which the eclipse occurs, then select the appropriate eclipse from the drop-down menu. Next, enter your location (latitude and longitude). See Sectin 3.4.6.4 (Transit of Venus) on page 45.

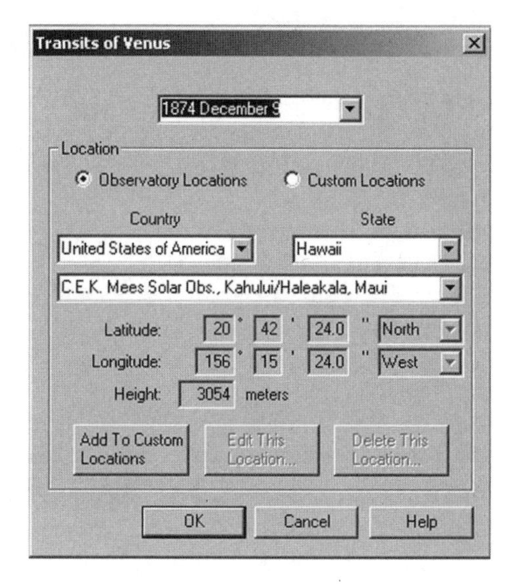

4.27 Time Systems

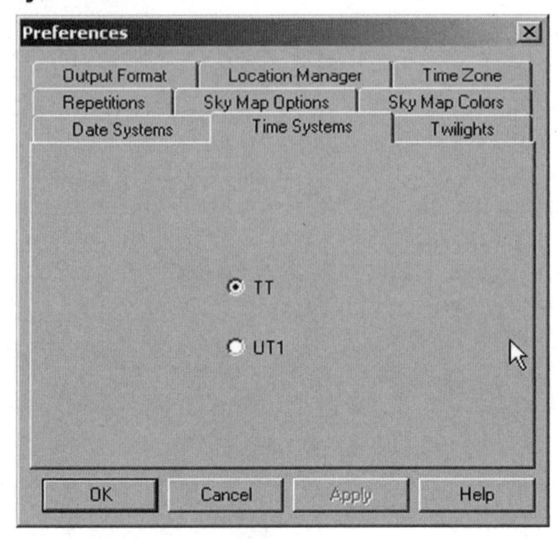

Time systems refers to which astronomical time scale you wish to use in your calculations. TT is terrestrial time, which is the scale used in the ephemerides used in MICA 2.0. UT1 is the same as TT, except that it is corrected for ΔT.

4.28 Time Zones

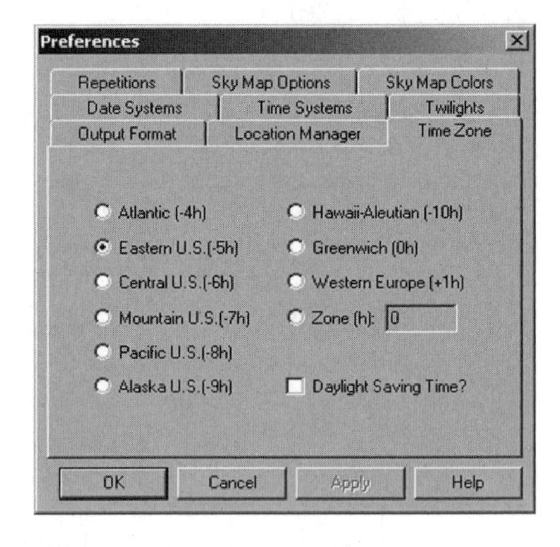

If you are about to perform a rise/set or sky map calculation, you can opt to use zone time. You cannot use zone time for any other calculation, because zone time is an offset of UTC, which is not used by MICA 2.0 Rise/set and sky map, however, have output that is sufficiently inexact that the UTC offset of UT1 from UTC is irrelevant.

If the desired time zone is not listed explicitly, enter it manually by selecting the 'Zone (h)' radio button, and enter the value (negative is west of Greenwich).

If you wish to take daylight savings time into account, click the appropriate check box. Note that MICA 2.0 does not calculate when DST is in effect (as this varies with location); using this option may lead to misleading results. See Section 3.4.3 (Rise/Set/Transit) on page 27.

4.29 Twilights

When doing a sunrise/sunset calculation, one has the option of specifying which type of twilight one wishes to calculate. This is where you make that selection. See Section 3.4.3.2 (Twilight) on page 28.

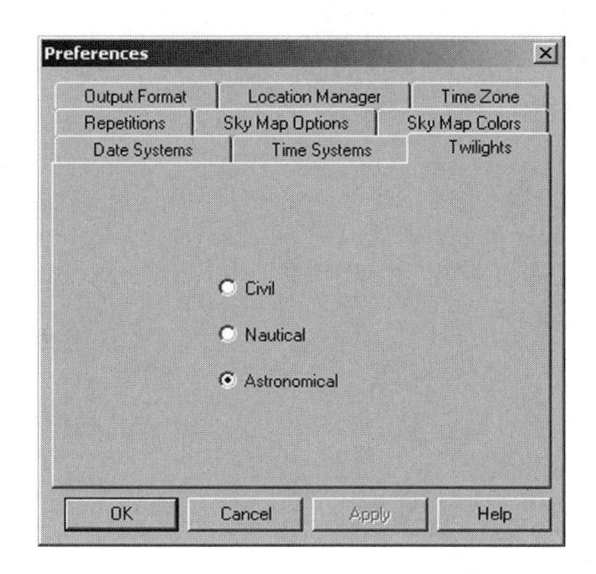

Chapter 5

MICA Dialog Boxes for the Mac

5.1 Asteroid Dialog Box

The Asteroid Dialog Box (*Compute|Positions|Asteroids*) lists the 15 asteroids for which MICA 2.0 has ephemerides. This dialog box view is used in the Position and Rise/Set/Transit Computations. Sections 3.4.1 (Positions) on page 19 and 3.4.3 (Rise/Set/Transit) on page 27.

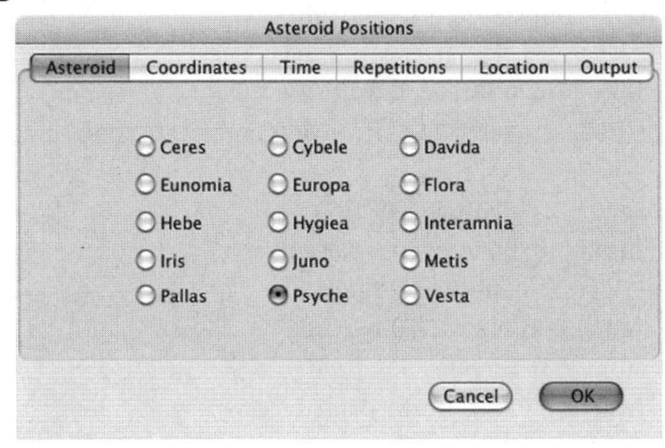

5.2 Bright Star Dialog Box

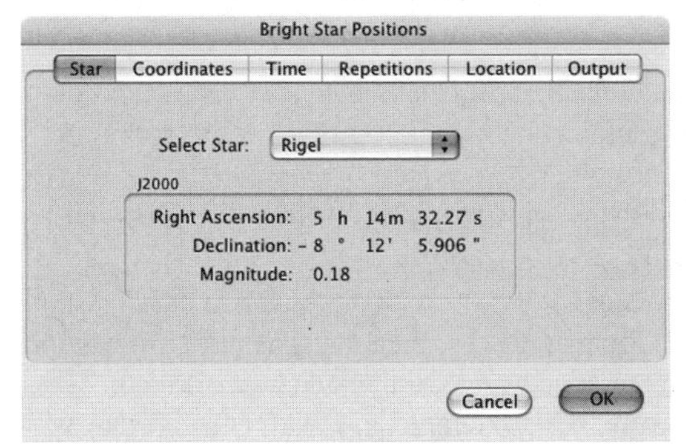

The Bright Star Dialog Box (*Compute|Positions|Bright Stars*) lists the 22 bright stars that can be used in MICA 2.0. This is the same list that was used in MICA 1.5. This dialog box view is used in the Position and Rise/Set/Transit Computations. See Sections 3.4.1 (Positions) on page 19 and 3.4.3 (Rise/Set/Transit) on page 27.

5.3 Catalog Search

Search Catalog(s): *(Compute|Rise/Set/Tranis|Catalog Object(s))* This Catalog selection dialog box view is used in the Position and Rise/Set/Transit Computations. Select the catalog(s) which are to be searched. Multiple catalogs may be selected. Several external catalogs have been included with the MICA 2.0 release. A user-supplied catalog may also be selected. The user-supplied catalog must follow the MICA 1.x ASCII format. See Sections 3.4.1 (Positions) on page 19 and 3.4.3 (Rise/Set/Transit) on page 27.

5.3.1 Catalog Search

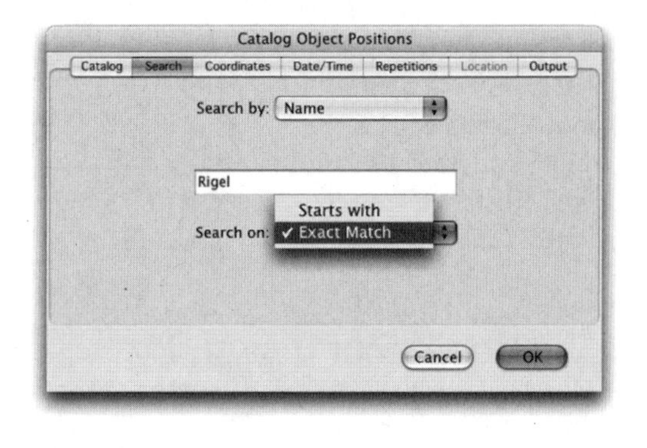

This dialog box view is used in the Position and Rise/Set/Transit computations. Select Name in the Search By field. This will search for a specific catalog identifier. Two types of searches can be performed:

- **Exact Match** This search will match the entered object name or number exactly. MICA will properly format the input object name before searching a given catalog. Extra blank white spaces are automatically removed or added, as appropriate, before performing the search. MICA object name searches are also case insensitive (i.e. the following searches are equivalent: 'Alf Ori', 'ALF ORI', and/or 'alf ori').
- **Starts with** This will match all objects with the specified initial characters and with the same spacing format as the catalog identifiers (i.e. white spaces are not removed or added before the search is performed). This search is also case insensitive. For example, 'alf' will return all stars with names that begin with 'alf' such as, 'alf Ori', or 'alf CMa'. Another example, 'HIP 2' will find nothing because the Hipparcos ID format is 'HIPbNNNNNN' and there are no Hipparcos ID's in the MICA catalog with ID's

5.3.2 Catalog Coordinate Search

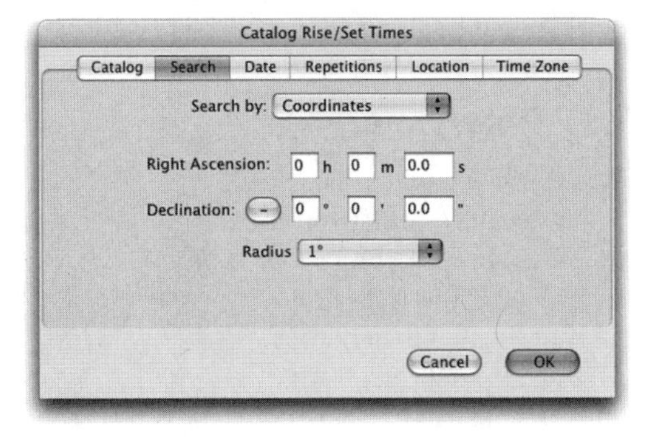

This dialog box view is used in the Position and Rise/Set/Transit computations. It is used to specify the search criteria when doing a catalog coordinate search. Select Coordinates in the Search By field. Specify the Right Ascension (in hours, minutes, seconds) and Declination (in degrees, arcminutes, arcseconds) of the center of the region (box) to be searched. The Radius (Half box widths) ranges from 15" to 5 degrees.

5.3.3 Catalog Visual Magnitude Search

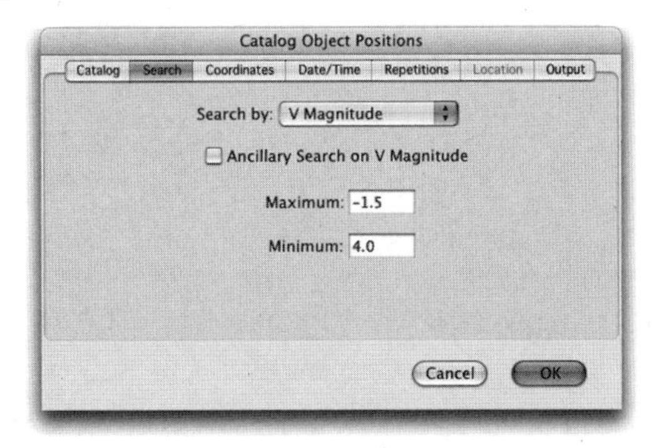

This dialog box view is used in the Position and Rise/Set/Transit computations and is used to specify the range of magnitudes to search in a catalog. To delete a custom location press the 'Delete' button. Select Visual Magnitude in the Search By field. Be sure that the checkbox is checked and specify the range of magnitudes to be searched. Note that the Visual magnitude search is done only in conjunction with either an object name or a coordinate range search. A Visual magnitude search of an entire catalog cannot be done with the Mac version of MICA 2.0.

5.4 General Preferences

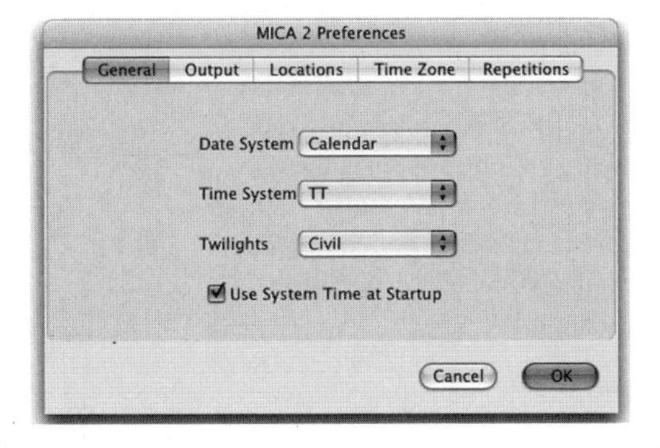

This dialog box view is used to control several MICA general preferences. The Date System (Calendar or Julian Date), the Time System (TT or UT1) and the Twilights (for Rise/Set/Transit only) type (Civil, Nautical or Astronomical) are set with this Preferences view. The box at the bottom should be checked if the computer system time will automatically be utilized for the date/time input field.

5.5 Date/Time

The Date/Time dialog box view is used to specify the date, time, and the time system (TT or UT1). This control is utilized in many of the MICA computations (Positions, Date & Time, Configurations, etc.)

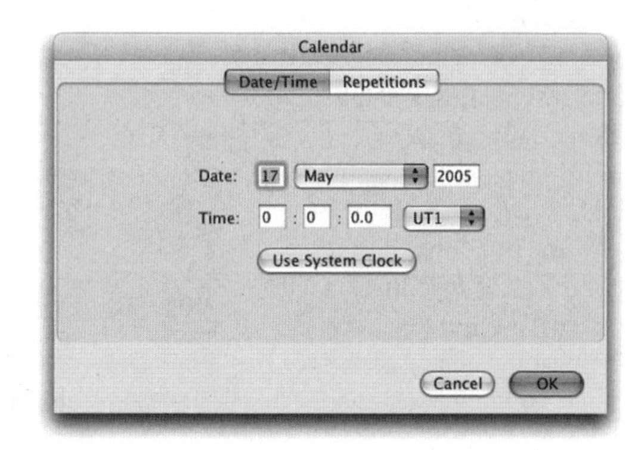

5.6 Galilean Moons

There are four calculation options to choose from:

- Position Differences (in RA, Dec, Dist, and PA)
- Astrometric Geocentric Equator of J2000.0
- Apparent Geocentric Equator of Date
- Topocentric Horizon

See Section 3.4.7 (Galilean Moons) on page 46.

5.7 Greatest Elongation

MICA will tabulate the times and dates of greatest elongations of Mercury and/or Venus from the Sun. This dialog box view is used to specify the objects (Mercury, Venus or both) for the calculation. See Section 3.4.8.5 (Greatest Elongations) on page 50.

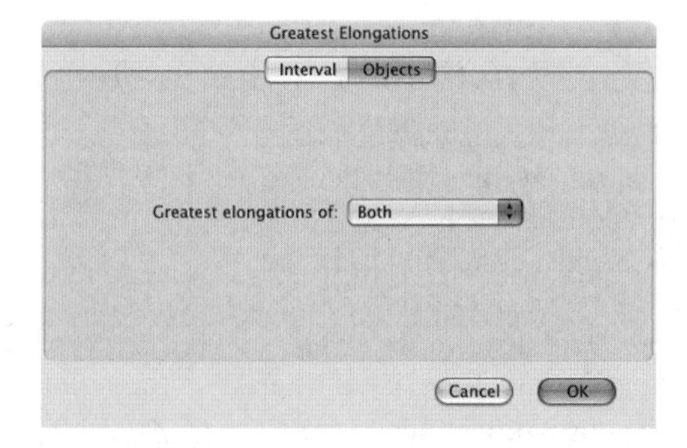

5.8 Location Manager

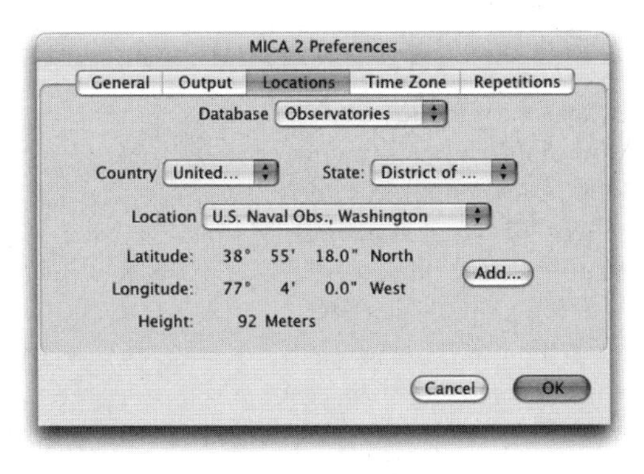

The location manager is used in any of the MICA computations involving topocentric calculations, including the Sky Map utility. The location manager consists of a pair of databases, one provided by the USNO, Observatory Locations, and one the user constructs, Custom Locations. The Observatory Database consists of a list 469 observatories worldwide. This is the same list that appears in The Astronomical Almanac. To select one of the observatories, select the Observatories Database, select the desired country and state (if the country selected is the United States) and the specific Observatory. If one wishes to add a particular observatory location to the Custom Locations list, simply click the 'Add..' button.

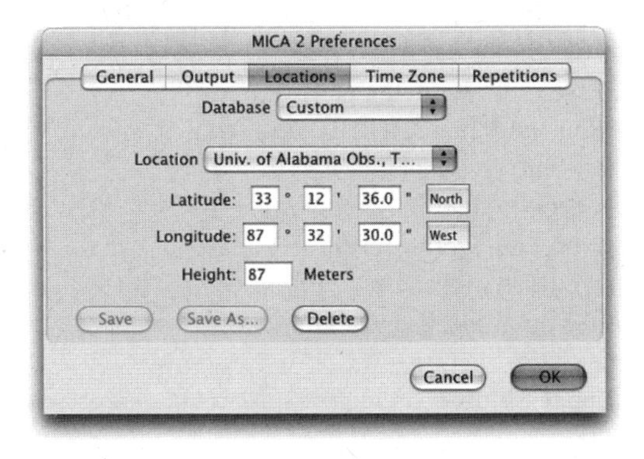

To utilize a custom location, select Custom from the Database list. The Location popup menu lists the locations stored in the database (if any). Location entries can be edited by simply altering the data for the location and clicking the 'Save' button. The new data for that location will then be entered into the Custom Locations database. To enter a new location, enter the latitude, longitude and height information. MICA will use the information as the default location until new location data is entered. However, the location will not be stored in the database until the 'Save as...' button is clicked. This will prompt the user to enter the location name.

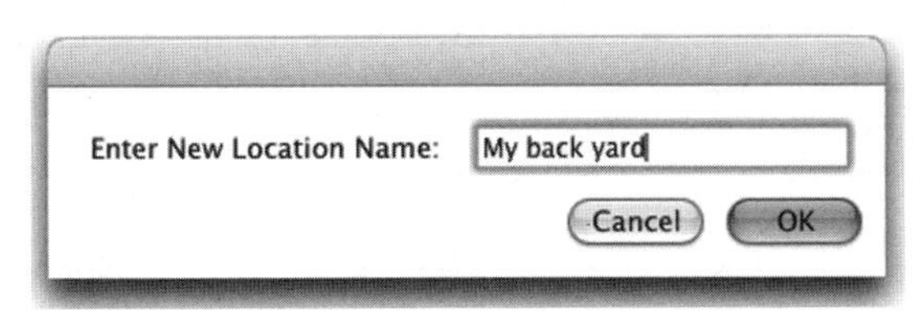

Once the location name is entered, click OK to add the location to the Custom Locations database.

5.9 Lunar Conjunctions

Lunar conjunctions are between the Earth, Moon, and a third, conjuncting body. This dialog box allows one to choose amongst a selection of all possible planets (the Earth excluded, of course), all possible asteroids (the one's for which we have ephemerides), and a small selection of stars. See Section 3.4.8.3 (Conjunctions) on page 49.

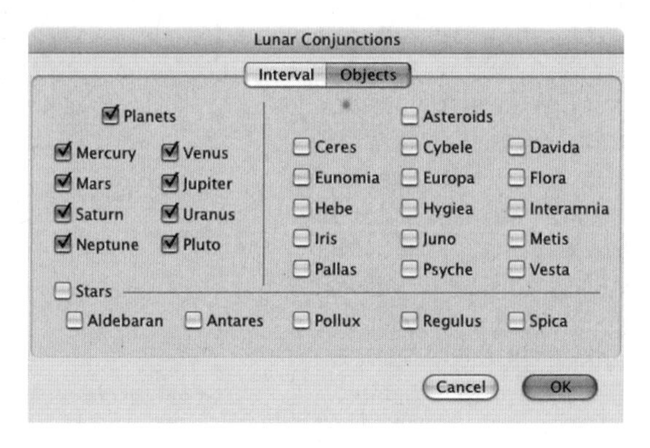

5.10 Lunar Eclipses

This dialog box is used calculate the local circumstances of a lunar eclipse. Enter the year in which the eclipse occurs and then select the appropriate eclipse from the drop-down menu. See the location manager description for information on how to enter the latitude and longitude. See Section 3.4.6.2 (Lunar Eclipse) on page 43.

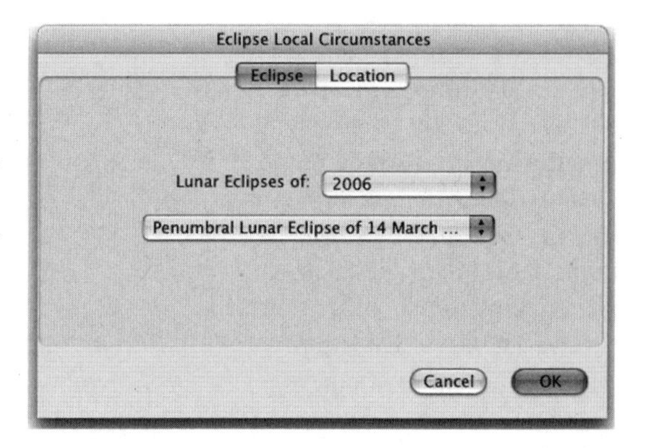

5.11 Moon Phases

This dialog box is used to specify the date range for a moon phase calculation. Note that the moon's phases are independent of the observer's location. See Section 3.4.8.2 (Moon Phases) on page 48.

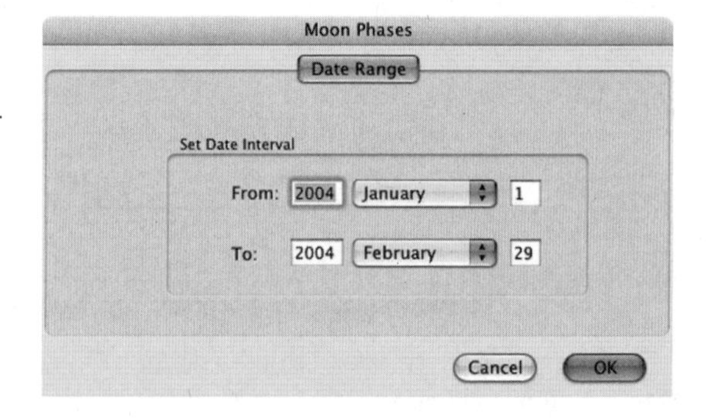

5.12 Output Format

You can specify the numerical output for most types of calculations in decimal format, or in DMS (degree, arc minutes, arc seconds) / HMS (hours, time minutes, time seconds).

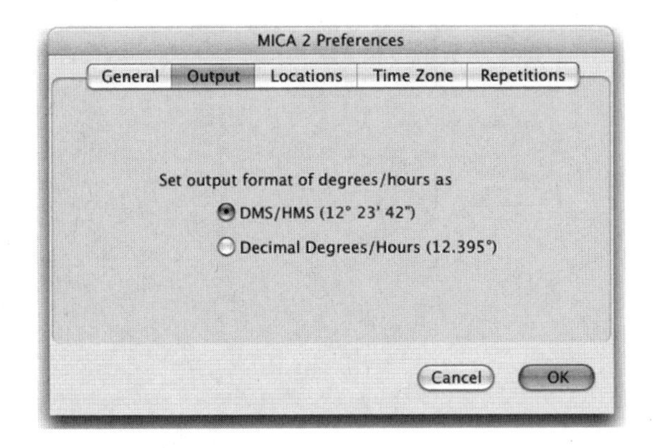

5.13 Oppositions

Oppositions between the Sun and other solar system objects can be calculated using this dialog box. The two check boxes at the top of the dialog box view will select all planets or all asteroids. See Section 3.4.8.4 (Oppositions) on page 50.

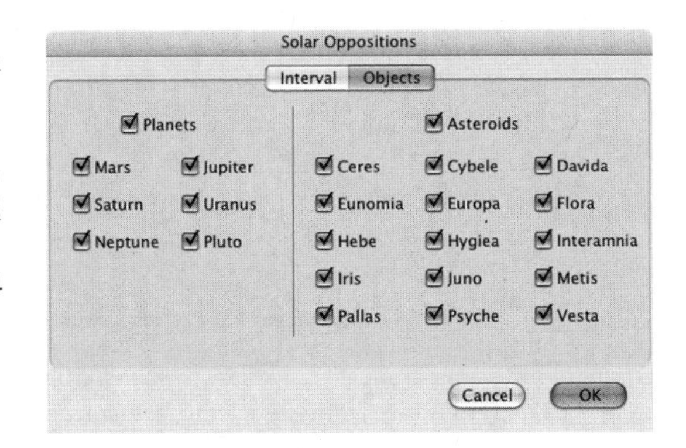

5.14 Phenomena Search

The Phenomena Search dialog box consists of 6 views for entry of the date Interval, Sun, Moon, Conjunctions, Oppositions, and Other information. The dialog box shows the Interval view which allows entry of the date range. At the bottom of the Phenomena Search dialog box are two buttons ('Search All' and 'Clear All'), which are used to select all possible search criteria and to clear all previously selected search criteria. See Section 3.4.8.6 (Phenomena Search) on page 50.

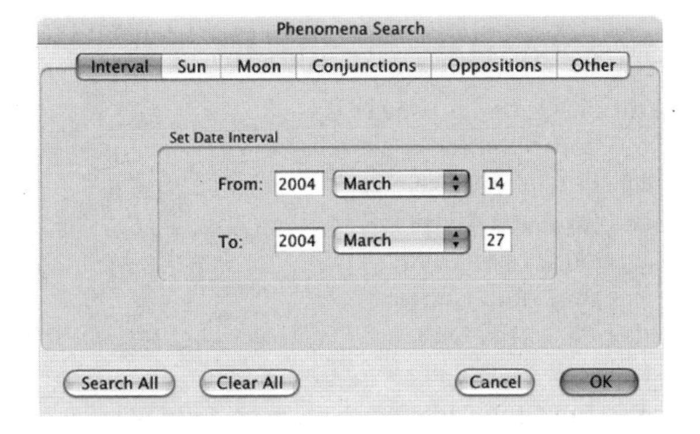

5.15 Physical Ephemeris

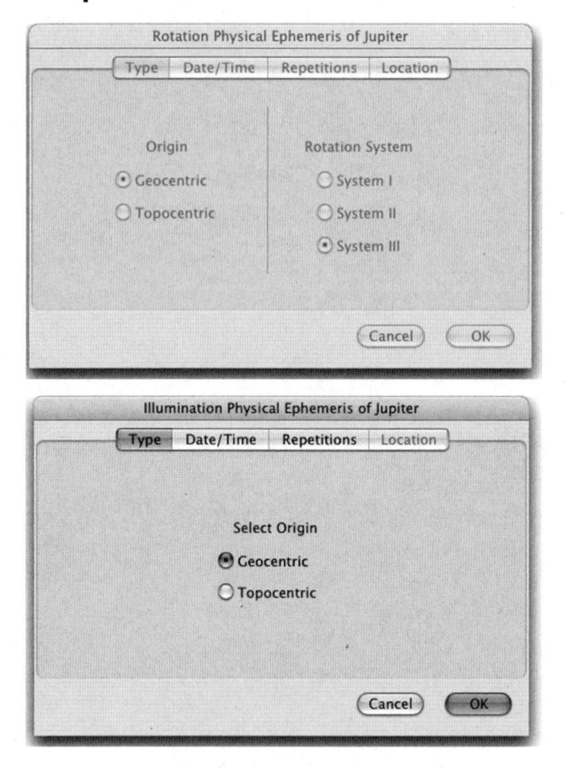

To do a physical ephemeris calculation, select either Illumination of Disc or Rotation of Disc from the Compute Menu then select an object, an origin, and a rotation system (possible only when doing a Jupiter rotation). See Section 3.4.4 (Physical Ephemeris) on page 31

5.16 Planetary Conjunctions

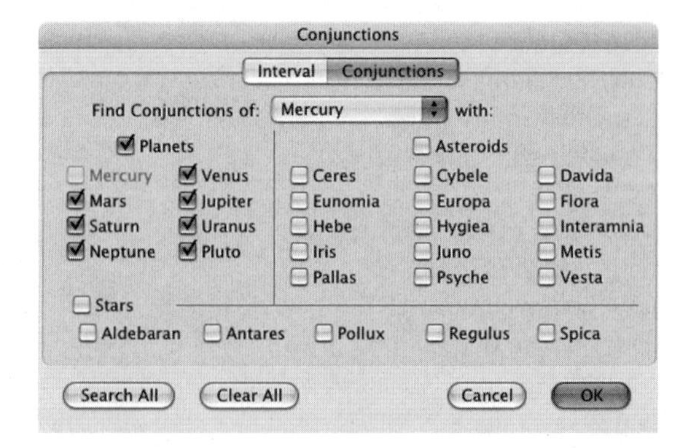

This dialog box is used to specify the objects (other planets, asteroids, selected stars) to search for planetary conjunctions with. Note that more than one combination of searches can be performed at the same time. For example, a planetary conjunction of Mercury with all other planets and a conjunction of Venus with all stars can searched for at the same time. Use the 'Search All' button to find all possible combinations of planetary conjunctions. Use the 'Clear All' button to clear the search criteria. See Section 3.4.8.3 (Conjunctions) on page 49.

5.17 Positions Coordinate System

This dialog box view is used to specify the Coordinate System or Position Types for a Position calculation. The allowed positions types depends upon which object is selected. More information on the Position types can be found in the Chapter on MICA Position Calculations. See Section 3.4.1 (Positions) on page 19.

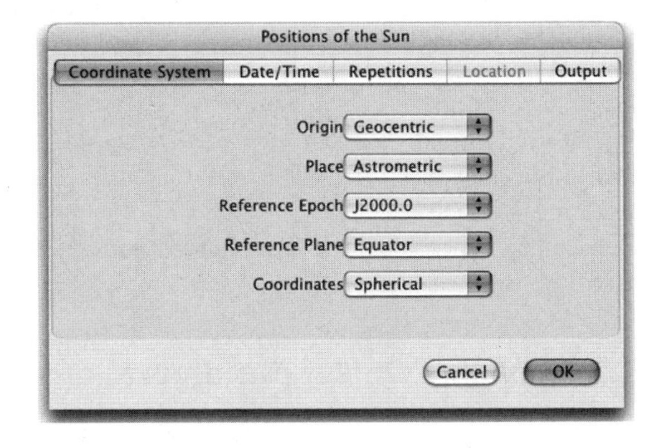

5.18 Repetitions

For many types of calculations, it is necessary to specify how many times you wish to perform the calculation, and over what time interval. This is where selections are made. Please note that, when doing rise/set calculations, the increment will be rounded to the nearest day (rounding up to one day if the interval was less than that).

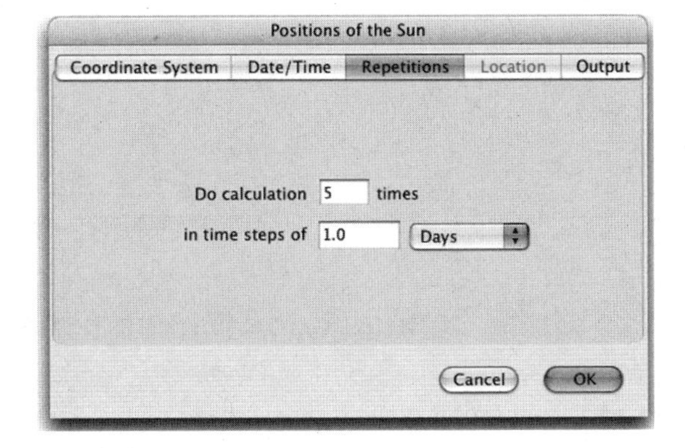

5.19 Sky Map Colors (Configurations)

The colors for the Sun, Moon, Planets, Asteroids, Stars, and Sky background can be individually selected by clicking on the color bar adjacent to the object. This will bring up a Macintosh utility to change the color. See Section 3.4.5.3 (Sky Map Configurations) on page 39.

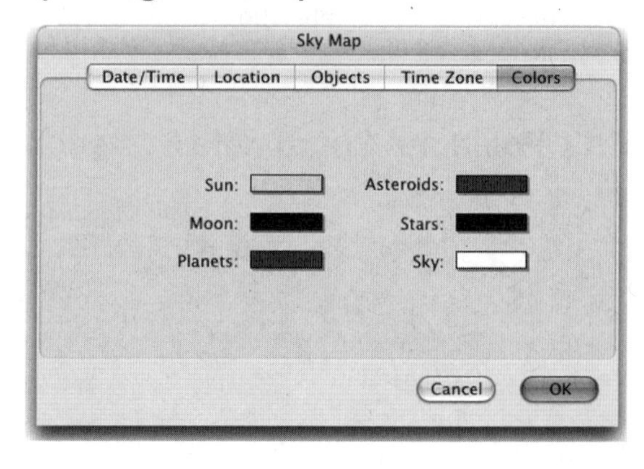

5.20 Sky Map Object (Configurations)

This dialog box view is used to specify which objects will be displayed in a Sky Map. Either all stars or a magnitude range may be specified. See Section 3.4.5.3 (Sky Map Configurations) on page 39.

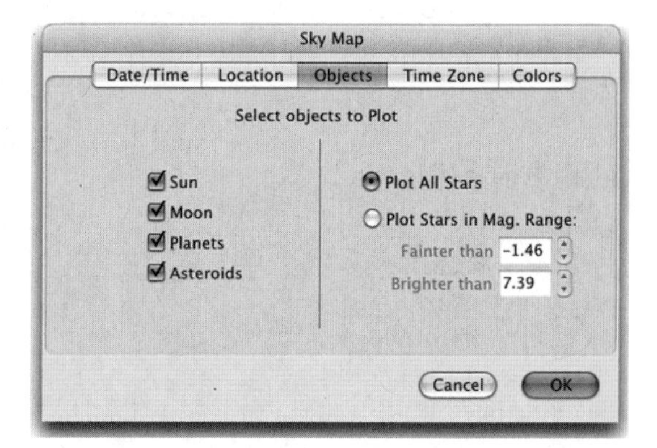

5.21 Solar Conjunctions (Phenomena)

Conjunctions between the Sun and other solar system objects can be calculated using this dialog box. One can select (or deselect) all planes or all asteroids by using the check boxes at the top of the dialog box view. See Section 3.4.8.3 (Conjunctions) on page 49.

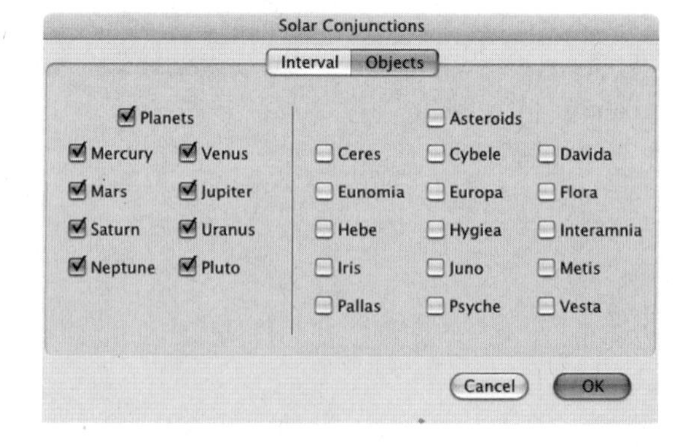

5.22 Solar Eclipse (Eclipses and Transits)

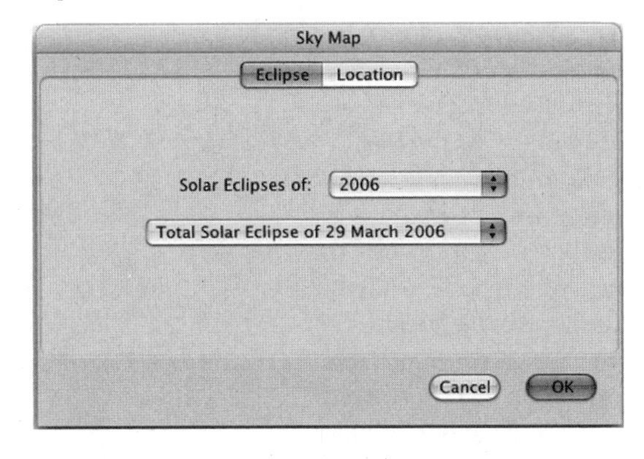

This dialog box is used to calculate the local circumstances of a solar eclipse. Enter the year in which the eclipse occurs and then select the appropriate eclipse from the drop-down menu. See the location manager description for information on how to enter the latitude and longitude.

To calculate the local circumstances of a solar eclipse, select Solar Eclipses from the MICA Eclipses and Transits sub-menu. Enter the year in which the eclipse occurs, then select the appropriate eclipse from the drop-down menu. Next, enter your location (latitude and longitude). The Position Angle of a given contact point on the solar limb is measured eastward (counterclockwise) around the solar limb, from the point on the Sun that is farthest north. See Section 3.4.6.1 (Solar Eclipse) on page 42.

5.23 Transit of Mercury (Eclipses and Transits)

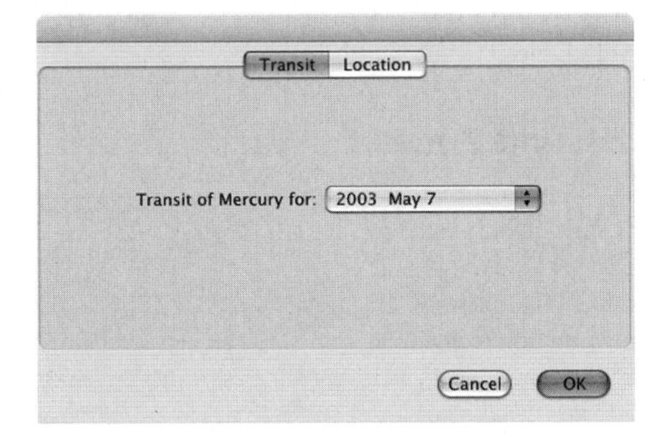

To calculate the local circumstances of a Mercury transit, select Transits of Mercury from the MICA Eclipses and Transits sub-menu. Use the above dialog box view to specify the date of a specific Mercury transit. See the location manager description for information on how to enter the latitude and longitude. See Section 3.4.6.3 (Transit of Mercury) on page 44.

5.24 Transit of Venus

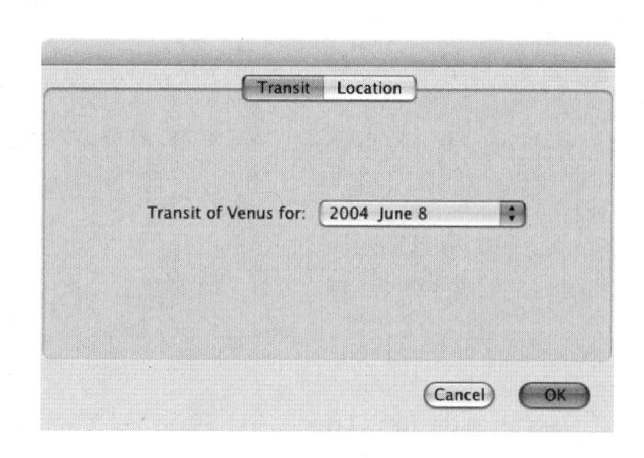

To calculate the local circumstances of a Mercury transit, select Transits of Mercury from the MICA Eclipses and Transits submenu. Use the above dialog box view to specify the date of a specific Mercury transit. See the location manager description for information on how to enter the latitude and longitude. See Sectin 3.4.6.4 (Transit of Venus) on page 45

5.25 Time Systems

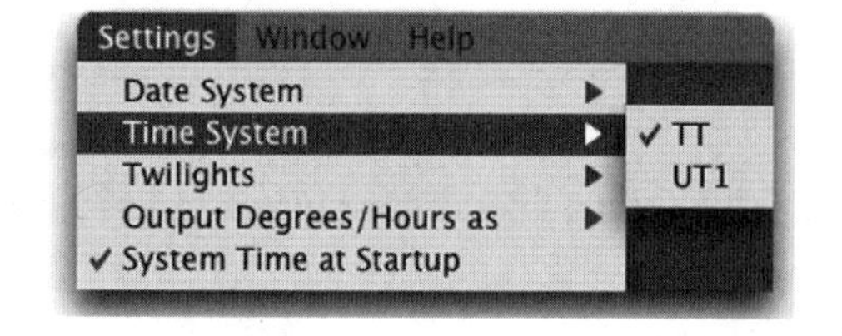

Time systems refers to which astronomical time scale you wish to use in your calculations. TT is terrestrial time, which is the scale used in the ephemerides used in MICA 2.0. UT1 is the same as TT, except that it is corrected for ΔT.

5.26 Time Zones

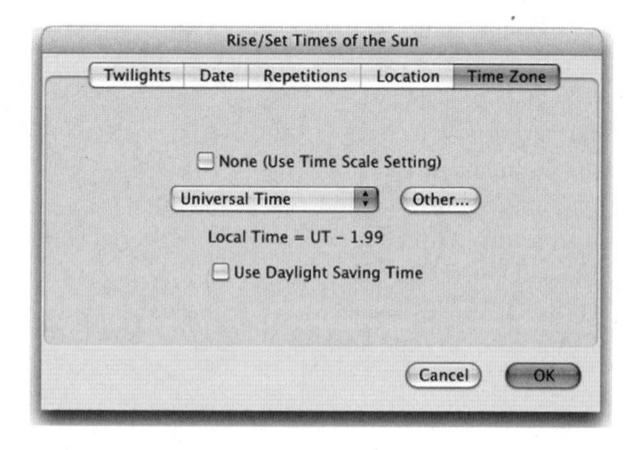

The Time Zone dialog box view is only used in the Rise/Set/Transit and the Sky Map utilities. If the None box is checked then the time scale (TT or UT1) specified elsewhere in the preferences will be utilized in the computations. If the desired time zone is not listed explicitly in the drop-down list, then it may be entered manually by pressing the 'Other…' button. Be sure to check the Use Daylight Saving Time box if you wish to take daylight savings time into account (it is not done automatically). See Section 3.4.3 (Rise/Set/Transit) on page 27.

5.27 Twilights (Rise/Set/Transit)

This dialog box view is used to specify the twilight type (Civil, Nautical or Astronomical) for a Rise/Set/Transit calculation for the Sun. See Section 3.4.3.2 (Twilight) on page 28.

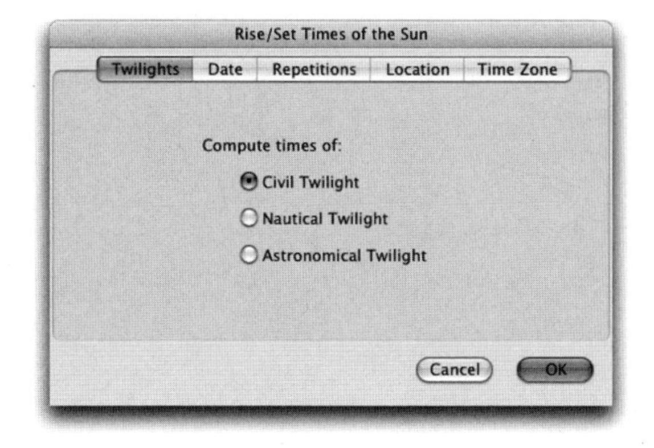

Appendix A Time Zones

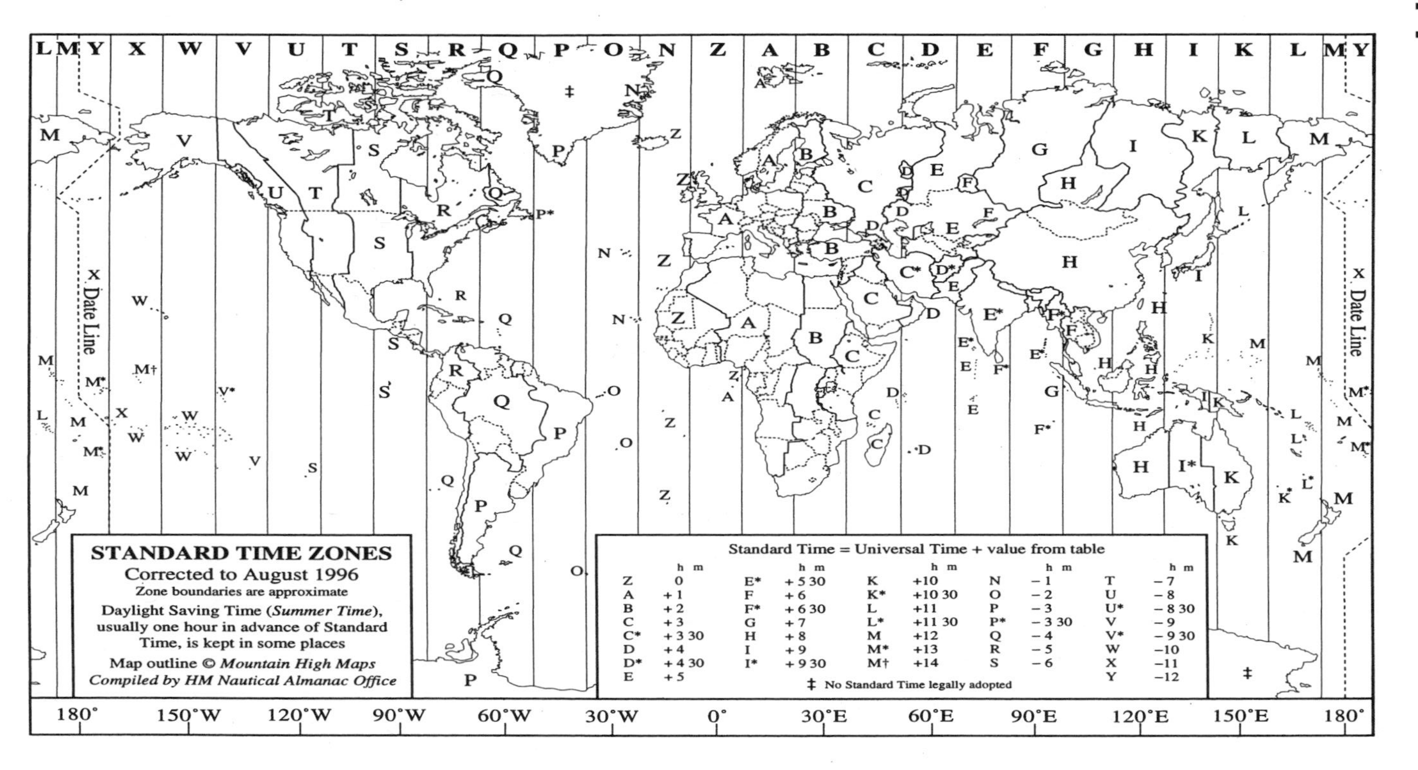

References

Argelander F., 1903, *Bonner Durchmusterung des nordlichen Himmels*, Bonn.

Cannon, A. J. and Pickering, E. C., 1918, *Ann. Astr. Obs. Harvard College*, **91**.

Cannon, A. J., 1925, "The Henry Draper Extension" (HDE), *Ann. Harvard Obs.*, **100**.

Cannon, A. J. and Mayall, M.W. , 1949, "The Henry Draper Extension II", **Ann. Harvard Obs.**, 112, 1–295.

Carrington, R. C., 1863, *Observations of the Spots on the Sun*, p. 244.

Fey, A.L., Ma, C., Arias, E.F., Charlot, P., Feissel-Vernier, M., Gontier, A.-M., Jacobs, C.S., Li, J., MacMillan, D.S., 2004, "The Second Extension of the International Celestial Reference Frame: ICRF-Ext.2,"*Astron J.*, **127**, pp. 3587–3608.

Fricke W., Schwan H., and Lederle T., 1988, "Fifth fundamental Catalogue," *Veroeff. Astron. Rechen-Inst. Heidelberg*, **32**, pp. 1–106.

Fricke W., et al., 1991, "Fifth fundamental Catalogue Part II. The FK5 extension", *Veroeff. Astron. Rechen-Inst. Heidelberg*, **33**, pp. 1.

Gill D, and Kapteyn J.C. 1896, "Cape photographic Durchmusterung for the equinox 1875 (–18 TO –37)," *Ann. Cape Obs.*, **3**, 1.

Gill D, and Kapteyn J.C. 1897, "Cape photographic Durchmusterung for the equinox 1875 (–38 TO –52)," *Ann. Cape Obs.*, **4**, 1.

Gill D, and Kapteyn J.C. 1900, "Cape photographic Durchmusterung for the equinox 1875 (–53 TO –89)," *Ann. Cape Obs.*, **5**, 1.

Gill D, 1903, "Revision of the Cape photographic Durchmusterung for the equinox 1875," *Ann. Cape Obs.*, **9**, 1.

Harris, D. L., 1961, in *Planets and Satellites* edited by G. P. Kuiper and B.A. Middlehurst, (chicago), pp. 272–342.

Hilton, J., 1992, "Physical Ephemerides of the Sun, Moon, Planets, and Satellites," in *The Explanatory Supplement to The Astronomical Almanac*, edited by Seidelmann, P. K., (University Science Books), pp. 383–419.

Hilton, J., 1999, "U.S. Naval Observatory Ephemerides of the Largest Asteroids", *Astron, J.*, **117**, pp. 1077–1086.

Hilton, J., 2003, "Updating the Visual Magnitudes of the Planets in *The Astronomical Almanac*, I. Mercury and Venus", AA Department internal memo.

The Hipparcos and Tycho Catalogues, 1997, **1–17**, European Space Agency SP-1200.

Hoffleit, et al. 1982, *The Bright Star Catalogue*.

Hohenkerk, C.Y. and Sinclair, A.T., 1985. *NAO Technical Note No. 63* (H.M. Nautical Almanac Office).

Hohenkerk, C.Y., Yallop, B.D., Smith, C.A., and Sinclair, A.T., 1992, "Celestial Reference Systems," in *The Explanatory Supplement to the Astronomical Almanac*, edited by Seidelmann, P. K., (University Science Books), p. 141 (section 3.281).

Johnson, T., 2004, private communication, USNO Earth Orientation Department.

Kaplan, G.H., Hughes, J. A., Seidelmann, P.K., Smith, C.A., and Yallop, B.D. 1989, "Mean and Apparent Place Computations in the New IAU System. III. Apparent, Topocentric, and Astrometric Places of Planets and Stars", *The Astronomical Journa*l, **97**, pp. 1197–1210.

Keenan, P. C. and McNeil, R. C., 1976, *Atlas of Spectra of the Cooler Stars: Types G, K, M, S, and C.*, Ohio State University Press.

Kukarin, et al. 1971, "General Catalog of Variable Stars, 3rd Ed.," published by Moscow, Academy of Sciences of the USSR, 4 editions and supplements (1971–1990).

Lieske, J. H. 1977b, *JPL Engineering Memorandum*, **314–112**.

Lieske, J. H. 1977a, Astron. Astrophys., **56**, 333.

Ma, C., Arias, E.F., Eubanks, T.M., Fey, A.L., Gontier, A.M., Jacobs, C.S., Sovers, O.J., Archinal, B.A., Charlot, P., 1997, "Definition and Realization of the International Celestial Reference System by VLBI Astrometry of Extragalactic Object," *International Earth rotation Service (IERS) Technical Note*, **23**, (Observatoire de Paris).

McCarthy, D. D. and Babcock, A. K., 1986, "The length of day since 1656," *Physics of the Earth and Planetary Interiors*, **44**, p. 281.

Morgan, W. W., H.A. Abt, and Tapschott, J.W., 1978, *Revised MK Spectral Atlas for Stars Earlier Than the Sun*, (Yerkes Obs. and Kitt Peak Nat. Obs.).

Perrine 1932, "Cordoba Durchmusterung declination –62 to –90," *Results of National Argentine Obs.*, **21**, pp. 1–140.

SAO Staff, 1962, *SAO Star Catalogue*, Publ. of the Smithsonian Institution of Washington, D.C. **4652**.

Schoenfeld E., 1886, *Bonner Durchmusterung des sudlichen Himmels*, Bonn .

Seidelmann, P.K, (editor), 1992, *Explanatory Supplement to The Astronomical Almanac*, University Science Books.

Seidelmann, P.K, et al, 2002, "Report of the IAU/IAG Working Group on Cartographic Coordinates and Rotation Elements of the Planets and Satellites: 2000," *Cel. Mech.*, **82**, Issue 1, pp. 83–111.

Smith, C.A., Kaplan, G.H., Huges, J.A., Seidelmann, P.K., Yallop, B.D., and Hohenkerk, C.Y., 1989, "Mean and apparent place computations in the new IAU system. I. The transformation of astrometric catalog systems to the equinox J2000.0.,", *Astron, J.*, **97**, pp. 265–273.

Standish, E. M., 1998, "JPL Planetary and Lunar Ephemerides, DE405/LE405," JPL *Interoffice Memorandum, IOM 312.F-98-048.*

Thome, J.M. 1892, "Cordoba Durchmusterung declination –22 to –32," *Results of National Argentine Obs.*, **16**, pp. 1–604.

Thome, J.M. 1894, "Cordoba Durchmusterung declination –32 to –42," *Results of National Argentine Obs.*, **17**, pp. 1–305.

Thome, J.M. 1900, "Cordoba Durchmusterung declination –42 to –52," *Results of National Argentine Obs.*, **16**, pp. 1–502.

Thome, J.M. 1914, "Cordoba Durchmusterung declination –52 to –62," *Results of National Argentine Obs.*, **21**, pp. 1–305.

Yallop, B.D., Hohenkerk, C.Y., Smith, C.A., Kaplan, G.H., Hughes, J.A., and Seidelmann, P.K., 1989, "Mean and apparent place computations in the new IAU system. II. Transformation of mean star places from FK4 B1950.0 FK5J2000.0 using matrices in 6-space", *Astron, J.*, **97**, pp. 274–279.

Glossary

ΔT

The difference between Terrestrial Time and Universal Time (UT): $\Delta T = TT - UT1$.

ΔUT1

The value of the difference between Universal Time (UT) and Coordinated Universal Time: $\Delta UT1 = UT1 - UTC$.

aberration

The apparent angular displacement of the observed position of a celestial object from its geometric position, caused by the finite velocity of light in combination with the motions of the observer and of the observed object. (See aberration, planetary.)

aberration, annual

The component of stellar aberration resulting from the motion of the Earth about the Sun. (See aberration, stellar.)

aberration, diurnal

The component of stellar aberration resulting from the observer's diurnal motion about the center of the Earth. (See aberration, stellar.)

aberration, E-terms of

The terms of annual aberration which depend on the eccentricity and longitude of perihelion. (See perihelion; aberration, annual).

aberration, elliptic

See aberration, E-terms of.

aberration, planetary

The apparent angular displacement of the observed position of a solar system body produced by motion of the observer and the actual motion of the observed object. (See aberration, stellar.)

aberration, secular

The component of stellar aberration resulting from the essentially uniform and almost rectilinear motion of the entire solar system in space. Secular aberration is usually disregarded.

aberration, stellar

The apparent angular displacement of the observed position of a celestial body resulting from the motion of the observer. Stellar aberration is divided into diurnal, annual, and secular components. (See aberration, diurnal; aberration, annual; aberration, secular).

altitude

The angular distance of a celestial body above or below the horizon, measured along the great circle passing through the body and the zenith. Altitude is 90° minus zenith distance.

anomaly

The angular measurement of a body in its orbit from its perihelion. (See eccentric anomaly; true anomaly.)

aphelion

The most distance point from the Sun in a heliocentric orbit.

apogee

The point at which a body in orbit around the Earth reaches its farthest distance from the Earth. Apogee is sometimes used with reference to the apparent orbit of the Sun around the Earth.

apparent place

The coordinates of a celestial object at a specific date, obtained by removing from the directly observed position of the object the effects that depend on the topocentric location of the observer; i.e., refraction, diurnal aberration, and geocentric (diurnal) parallax. Thus, the position at which the object would actually be seen from the center of the Earth,— if the Earth were transparent, nonrefracting, and massless—referred to the true equator and equinox. (See aberration, diurnal.)

apparent solar time (AST)

The measure of time based on the diurnal motion of the true Sun. The rate of diurnal motion undergoes seasonal variation caused by the obliquity of the ecliptic and by the eccentricity of the Earth's orbit. Additional small variations result from irregularities in the rotation of the Earth on its axis.

argument of the pericenter

An angle measured within the orbit plane from the line of nodes toward the pericenter. The argument of the pericenter is one of the Keplerian elements parameterizing the orbit.

aspect

The apparent position of any of the planets or the Moon relative to the Sun, as seen from Earth.

astrometric ephemeris

An ephemeris of a solar system body in which the tabulated positions are essentially comparable to catalog mean places of stars at a standard epoch. An astrometric position is obtained by adding to the geometric position, computed from gravitational theory, the correction for light-time. Prior to 1984, the E-terms of annual aberration were also added to the geometric position. (See aberration, annual; aberration E-terms of.)

astronomical coordinates

The longitude and latitude of a point on the Earth relative to the geoid. These coordinates are influenced by local gravity anomalies. (See zenith; longitude, terrestrial; latitude, terrestrial.)

astronomical unit (a.u.)

The radius of a circular orbit in which a body of negligible mass, and free of perturbations, would revolve around the Sun in $2\pi/k$ days, k being the Gaussian gravitational constant. This is slightly less than the semimajor axis of the Earth's orbit.

atomic second

See second, Système International.

augmentation

The amount by which the apparent semidiameter of a celestial body, as observed from the surface of the Earth, is greater than the semidiameter that would be observed from the center of the Earth.

azimuth

the angular distance measured clockwise along the horizon from a specified reference point (usually north) to the intersection with the great circle drawn from the zenith through a body on the celestial sphere.

barycenter

The center of mass of a system of bodies; e.g., the center of mass of the solar system or the Earth-Moon system.

barycentric

With reference to, or pertaining to, the barycenter of the solar system.

Barycentric Dynamical Time (TDB)

A time scale defined by an IAU 1976 resolution for use as an independent argument of barycentric ephemerides and equations of motion. TDB was defined to have only periodic variations with respect to what is now called Terrestrial Time (TT).

brilliancy

For Mercury and Venus the quantity ks^2/r^2, where $k = 0.5(1 + \cos i)$, i is the phase angle, s is the apparent semidiameter, r is the heliocentric distance.

calendar

A system of reckoning time in which days are enumerated according to their position in cyclic patterns.

catalog equinox

The intersection of the hour circle of zero right ascension of a star catalog with the celestial equator. (See dynamical equinox; equator).

celestial ephemeris pole

The reference pole for nutation and polar motion; the axis of figure for the mean surface of a model Earth in which the free motion has zero amplitude. This pole has no nearly-diurnal nutation with respect to a space-fixed or Earth-fixed coordinate system.

celestial equator

The plane perpendicular to the celestial ephemeris pole. Colloquially, the projection onto the celestial sphere of the Earth's equator. (See mean equator and equinox; true equator and equinox.)

celestial pole

Either of the two points projected onto the celestial sphere by the extension of the Earth's axis of rotation to infinity.

celestial sphere

An imaginary sphere of arbitrary radius upon which celestial bodies may be considered to be located. As circumstances require, the celestial sphere may be centered at the observer, at the Earth's center, or at any other location.

center of figure

That point so situated relative to the apparent figure of a body that any line drawn through it divides the figure into two parts having equal apparent areas. If the body is oddly shaped, the center of figure may lie outside the figure itself.

center of light

Same as center of figure except referring only to the illuminated portion.

conjunction

The phenomenon in which two bodies have the same apparent celestial longitude or right ascension as viewed from a third body. Conjunctions are usually tabulated as geocentric phenomena. For Mercury and Venus, geocentric inferior conjunctions occur when the planet is between the Earth and Sun, and superior conjunctions occur when the Sun is between the planet and Earth.

Coordinated Universal Time (UTC)

The time scale available from broadcast time signals. UTC differs from TAI by an integral number of seconds; it is maintained within +/–0.90 second of UT1 by the introduction of one second steps (leap seconds). (See International Atomic Time; Universal Time; leap second.)

culmination

Passage of a celestial object across the observer's meridian; also called "meridian passage." More precisely, culmination is the passage through the point of greatest altitude in the diurnal path. Upper culmination (also called "culmination above pole" for circumpolar stars and the Moon) or transit is the crossing closer to the observer's zenith. Lower culmination (also called "culmination below pole" for circumpolar stars and the Moon) is the crossing farther from the zenith.

day

An interval of 86 400 SI seconds, unless otherwise indicated. (See second, Système International.)

declination

Angular distance on the celestial sphere north or south of the celestial equator. It is measured along the hour circle passing through the celestial object. Declination is usually given in combination with right ascension or hour angle.

defect of illumination

The angular amount of the observed lunar or planetary disk that is not illuminated to an observer on the Earth.

deflection of light

The angle by which the direction of a light ray is altered from a straight line by the gravitational field of the Sun or other massive object. As seen from the Earth, objects appear to be deflected radially away from the Sun by up to $1''.75$ at the Sun's limb. Correction for this effect, which is independent of wavelength, is included in the reduction from mean place to apparent place.

deflection of the vertical

The angle between the astronomical vertical and the geodetic vertical. (See zenith; astronomical coordinates; geodetic coordinates.)

delta T

See ΔT.

delta UT1

See $\Delta UT1$.

direct motion

For orbital motion in the solar system, motion that is counterclockwise in the orbit as seen from the north pole of the ecliptic; for an object observed on the celestial sphere, motion that is from west to east, resulting from the relative motion of the object and the Earth.

diurnal motion

The apparent daily motion caused by the Earth's rotation, of celestial bodies across the sky from east to west.

dynamical equinox

The ascending node of the Earth's mean orbit on the Earth's true equator; i.e., the intersection of the ecliptic with the celestial equator at which the Sun's declination is changing from south to north. (See catalog equinox; equinox; true equator and equinox.)

dynamical time

The family of time scales introduced in 1984 to replace ephemeris time as the independent argument of dynamical theories and ephemerides. (See Barycentric Dynamical Time; Terrestrial Time.)

eccentric anomaly

In undisturbed elliptic motion, the angle measured at the center of the orbit ellipse from pericenter to the point on the circumscribing auxiliary circle from which a perpendicular to the major axis would intersect the orbiting body. (See mean anomaly; true anomaly.)

eccentricity

A parameter that specifies the shape of a conic section; one of the standard elements used to describe an elliptic or hyperbolic orbit. (See orbital elements.)

eclipse

The obscuration of a celestial body caused by its passage through the shadow cast by another body.

eclipse, annular

A solar eclipse in which the solar disk is never completely covered but is seen as an annulus or ring at maximum eclipse. An annular eclipse occurs when the apparent disk of the Moon is smaller than that of the Sun. (See eclipse, solar)

eclipse, lunar

An eclipse in which the Moon passes through the shadow cast by the Earth. The eclipse may be total (the Moon passing completely through the Earth's umbra at maximum eclipse), or partial (the Moon passing partially through the Earth's umbra at maximum eclipse), or penumbral (the Moon passing only through the Earth's penumbra).

eclipse, solar

An eclipse in which the Earth passes through the shadow cast by the Moon. It may be total (observer in the Moon's umbra), partial (observer in the Moon's penumbra), or annular. (See eclipse, annular.)

ecliptic

The mean plane of the Earth's orbit around the Sun.

elements, Besselian

Quantities tabulated for the calculation of accurate predictions of an eclipse or occultation for any point on or above the Earth.

elements, Keplerian

See Keplerian elements.

elements, mean

See mean elements.

elements, orbital

See orbital elements.

elements, osculating

See osculating elements.

elongation, greatest

The instant when the geocentric angular distance of Mercury or Venus from the Sun is at a maximum.

elongation, planetary

The geocentric angle between a planet and the Sun. Planetary elongations are measured from 0° to 180°, east or west of the Sun.

elongation, satellite

The geocentric angle between a satellite and its primary. Satellite elongations are measured from 0° east or west of the planet.

epact

The age of the Moon; the number of days since new Moon, diminished by one day, on January 1 in the Gregorian ecclesiastical lunar cycle. (See Gregorian Calendar; lunar phases.)

ephemeris

A tabulation of the positions of a celestial object in an orderly sequence for a number of dates.

ephemeris hour angle

An hour angle referred to the ephemeris meridian.

ephemeris longitude

Longitude measured eastward from the ephemeris meridian. (See longitude, terrestrial.)

ephemeris meridian

A fictitious meridian that rotates independently of the Earth at the uniform rate implicitly defined by Terrestrial Time (TT). The ephemeris meridian is 1.002 738 ΔT east of the Greenwich meridian, where $\Delta T = TT - UT1$.

ephemeris time (ET)

The time scale used prior to 1984 as the independent variable in gravitational theories of the solar system. In 1984, ET was replaced by dynamical time.

ephemeris transit

The passage of a celestial body or point across the ephemeris meridian.

epoch

An arbitrary fixed instant of time or date used as a chronological reference datum for calendars, celestial reference systems, star catalogs, or orbital motions. (See calendar; orbit).

equation of the equinoxes

The difference apparent sidereal time minus mean sidereal time, due to the effect of nutation on the location of the equinox. (See sidereal time.)

equation of time

The difference apparent solar time minus mean solar time.

equator

The great circle on the surface of a body formed by the intersection of the surface with the plane passing through the center of the body perpendicular to the axis of rotation. (See celestial equator.)

equinox

Either of the two points on the celestial sphere at which the ecliptic intersects the celestial equator; also the time at which the Sun passes through either of these intersection points; i.e., when the apparent longitude (see apparent place; longitude, celestial) of the Sun is 0° or 180°. (See apparent place; longitude, celestial; catalog equinox; dynamical equinox.)

era

A system of chronological notation reckoned from a given date.

flattening

A parameter that specifies the degree by which a planet's figure differs from that of a sphere; the ratio $f = (a - b)/a$, where a is the equatorial radius and b is the polar radius.

frequency

The number of cycles or complete alternations, per unit time, of a carrier wave, band, or oscillation.

frequency standard

A generator whose output is used as a precise frequency reference; a primary frequency standard is one whose frequency corresponds to the adopted definition of the second, with its specified accuracy achieved without calibration of the device. (See second, Système International.)

Gaussian gravitational constant (k = 0.017 202 098 95)

The constant defining the astronomical system of units of length (astronomical unit), mass (solar mass) and time (day), by means of Kepler's third law. The dimensions of k^2 are those of Newton's constant of gravitation $L^3M^{-1}T^{-2}$.

geocentric

With reference to, or pertaining to, the center of the Earth.

geocentric coordinates

The latitude and longitude of a point on the Earth's surface relative to the center of the Earth; also celestial coordinates given with respect to the center of the Earth. (See zenith; latitude, terrestrial; longitude, terrestrial.)

geodetic coordinates

The latitude and longitude of a point on the Earth's surface determined from the geodetic vertical (normal to the reference ellipsoid). (See zenith; latitude, terrestrial; longitude, terrestrial.)

geoid

An equipotential surface that coincides with mean sea level in the open ocean. On land it is the level surface that would be assumed by water in an imaginary network of frictionless channels connected to the ocean.

geometric position

The position of an object defined by a straight line (vector) between the center of the Earth (or the observer) and the object at a given time, without any corrections for light time, aberration, etc.

Greenwich Apparent Sidereal Time (GAST)

The Greenwich hour angle of the true equinox of date.

Greenwich Mean Sidereal Time (GMST)

The Greenwich hour angle of the mean equinox of date.

Greenwich Sidereal Date (GSD)

The number of sidereal days elapsed at Greenwich since the beginning of the Greenwich sidereal day that was in progress at the Julian date (JD) 0.0.

Greenwich sidereal day number

The integral part of the Greenwich sidereal date (GSD).

Gregorian calendar

The calendar introduced by Pope Gregory XIII in 1582 to replace the Julian calendar; the calendar now used as the civil calendar in most countries. Every year that is exactly divisible by four is a leap year, except for centurial years, which must be exactly divisible by 400 to be leap years. Thus, 2000 was a leap year, but 1900 and 2100 are not leap years.

height

Elevation above ground or distance upwards from a given level (especially sea level) to a fixed point. (See altitude.)

heliocentric

With reference to, or pertaining to, the center of the Sun.

horizon

A plane perpendicular to the line from an observer to the zenith. The great circle formed by the intersection of the celestial sphere with a plane perpendicular to the line from an observer to the zenith is called the astronomical horizon.

horizontal parallax

The difference between the topocentric and geocentric positions of an object, when the object is on the astronomical horizon.

hour angle

The angular distance on the celestial sphere measured westward along the celestial equator from the meridian to the hour circle that passes through a celestial object.

hour circle

A great circle on the celestial sphere that passes through the celestial poles and is therefore perpendicular to the celestial equator.

inclination

The angle between two planes or their poles; usually the angle between an orbital plane and a reference plane; one of the standard orbital elements that specifies the orientation of the orbit. (See orbital elements.)

instantaneous orbit

The unperturbed two-body orbit that a body would follow if perturbations were to cease instantaneously. Each orbit in the solar system (and, more generally, in the many-body setting) can be represented as a sequence of instantaneous ellipses or hyperbolae whose parameters are called orbital elements. If these elements are chosen to be osculating, each instantaneous orbit is tangential to the physical orbit. (See orbital elements; osculating elements.)

International Astronomical Union (IAU)

An international non-governmental organization that promotes the science of astronomy in all its aspects. The IAU is composed of both national and individual members. In the field of positional astronomy, the IAU, among other activities, recommends standards for data analysis and modeling, usually in the form of resolutions passed at General Assemblies held every three years.

International Atomic Time (TAI)

The continuous scale resulting from analyses by the Bureau International des Poids et Mesures of atomic time standards in many countries. The fundamental unit of TAI is the SI second on the geoid, and the epoch is 1958 January 1. (See second, Système International.)

International Celestial Reference Frame (ICRF)

The coordinates of 212 extragalactic radio sources that serve as fiducial points to fix the axes of the International Celestial Reference System (ICRS), recommended by the IAU in 1997.

International Celestial Reference System (ICRS)

A time-independent, kinematically non-rotating barycentric reference system recommended by the IAU in 1997. Its axes are those of the International Celestial Reference Frame (ICRF).

invariable plane

The plane through the center of mass of the solar system perpendicular to the angular momentum vector of the solar system.

irradiation

An optical effect of contrast that makes bright objects viewed against a dark background appear to be larger than they really are.

Julian calendar

The calendar introduced by Julius Caesar in 46 B.C. to replace the Roman calendar. In the Julian calendar a common year is defined to comprise 365 days, and every fourth year is a leap year comprising 366 days. The Julian calendar was superseded by the Gregorian calendar.

Julian date (JD)

The interval of time in days and fraction of a day since 4713 B.C. January 1, Greenwich noon, Julian proleptic calendar. In precise work, the timescale; e.g., Terrestrial Time (TT) or Universal Time (UT), should be specified.

Julian date, modified (MJD)

The Julian date minus 2400000.5 (17 November 1858).

Julian day number (JD)

The integral part of the Julian date (JD).

Julian proleptic calendar

The calendric system employing the rules of the Julian calendar, but extended and applied to dates preceding the introduction thereof.

Julian year

A period of 365.25 days. It served as the basis for the Julian calendar.

Keplerian Elements

A certain set of six orbital elements, sometimes referred to as the Keplerian set. Historically, this set included the mean anomaly at the epoch, the semimajor axis, the eccentricity and three Euler angles the longitude of the ascending node, the inclination, and the argument of pericenter. The time of pericenter passage is often used as part of the Keplerian set instead of the mean anomaly at the epoch. Sometimes the longitude of pericenter (which is the sum of the longitude of the ascending node and the argument of pericenter) is used instead of either the longitude of the ascending node or the argument of pericenter.

Laplacian plane

For planets see invariable plane; for a system of satellites, the fixed plane relative to which the vector sum of the disturbing forces has no orthogonal component.

latitude, celestial

Angular distance on the celestial sphere measured north or south of the ecliptic along the great circle passing through the poles of the ecliptic and the celestial object. Also referred to as ecliptic latitude.

latitude, terrestrial

Angular distance on the Earth measured north or south of the equator along the meridian of a geographic location.

leap second

A second added between 60s and 0s at announced times to keep UTC within $0\overset{s}{.}90$ of UT1. Generally, leap seconds are added at the end of June or December. (See second, Système International; Universal Time (UT); Coordinated Universal Time (UTC).)

librations

Variations in the orientation of the Moon's surface with respect to an observer on the Earth. Physical librations are due to variations in the orientation of the Moon's rotational axis in inertial space. The much larger optical librations are due to variations in the rate of the Moon's orbital motion, the obliquity of the Moon's equator to its orbital plane, and the diurnal changes of geometric perspective of an observer on the Earth's surface.

light, deflection of

See deflection of light.

light-time

The interval of time required for light to travel from a celestial body to the Earth. During this interval the motion of the body in space causes an angular displacement of its apparent place from its geometric place. (See geometric position; aberration, planetary.)

light-year

The distance that light traverses in a vacuum during one year.

limb

The apparent edge of the Sun, Moon, or a planet or any other celestial body with a detectable disc.

limb correction

A correction that must be made to the distance between the center of mass of the Moon and its limb. These corrections are due to the irregular surface of the Moon and are a function of the librations in longitude and latitude and the position angle from the central meridian.

local hour angle (LHA)

It is the local sidereal time minus the right ascension, LHA = LST – RA. The local hour angle is measured in units of time 1 hour for each 15 degrees from the local meridian. In MICA, hour angles are positive (0 to 12 hours) west and negative (0 to –12 hours) east of the local meridian.

local sidereal time

The local hour angle of a catalog equinox.

longitude, celestial

Angular distance on the celestial sphere measured eastward along the ecliptic from the dynamical equinox to the great circle passing through the poles of the ecliptic and the celestial object. Also referred to as ecliptic longitude.

longitude, terrestrial

Angular distance measured along the Earth's equator from the Greenwich meridian to the meridian of a geographic location.

longitude of the ascending node

Given an orbit and a reference plane through the primary body (or center of mass) the angle, O, at the primary, between a fiducial direction in the reference plane and the point at which the orbit crosses the reference plane from south to north. Equivalently, O is one of the angles in the reference plane between the fiducial direction and the line of nodes. It is one of the six Keplerian elements that specify an orbit. For planetary orbits, the primary is the Sun, the reference plane is usually the ecliptic, and the fiducial direction is usually toward the equinox. (See node; orbital elements; instantaneous orbit.)

luminosity-class

Distinctions in intrinsic brightness among stars of the same spectral type.

lunar phases

Cyclically recurring apparent forms of the Moon. New Moon, first quarter, full Moon and last quarter are defined as the times at which the excess of the apparent celestial longitude of the Moon over that of the Sun is 0°, 90°, 180° and 270°, respectively. (See longitude, celestial.)

lunation

The period of time between two consecutive new Moons.

magnitude, stellar

A measure on a logarithmic scale of the brightness of a celestial object considered as a point source.

magnitude of a lunar eclipse

The fraction of the lunar diameter obscured by the shadow of the Earth at the greatest phase of a lunar eclipse, measured along the common diameter. (See eclipse, lunar.)

magnitude of a solar eclipse

The fraction of the solar diameter obscured by the Moon at the greatest phase of a solar eclipse, measured along the common diameter. (See eclipse, solar.)

mean anomaly

The product of the mean motion of an orbiting body and the interval of time since the body passed the pericenter. Thus, the mean anomaly is the angle from the pericenter of a hypothetical body moving with a constant angular speed that is equal to the mean motion. In realistic computations, with disturbances taken into account, the mean anomaly is equal to its initial value at an epoch plus an integral of the mean motion over the time elapsed since the epoch. (See true anomaly; eccentric anomaly; mean anomaly at epoch.)

mean anomaly at epoch

The value of the mean anomaly at a specific epoch; i.e., at some fiducial moment of time. It is one of the six Keplerian elements that specify an orbit. (See Keplerian elements; orbital elements; instantaneous orbit.)

mean distance

An average distance between the primary and the secondary gravitating body. The meaning of the mean distance depends upon the chosen method of averaging (i.e., averaging over the time, or over the true anomaly, or the mean anomaly. It is also important what power of the distance is subject to averaging.) In this volume the mean distance is defined as the inverse of the time-averaged reciprocal distance $(\int r^{-1} dt)^{-1}$. In the two body setting, when the disturbances are neglected and the orbit is elliptic, this formula yields the semimajor axis, a, which plays the role of mean distance.

mean elements

Average values of the orbital elements over some section of the orbit or over some interval of time. They are interpreted as the elements of some reference (mean) orbit that approximates the actual one and, thus, may serve as the basis for calculating orbit perturbations. The values of mean elements depend upon the chosen method of averaging and upon the length of time over which the averaging is made.

mean equator and equinox

The celestial reference system defined by the orientation of the Earth's equatorial plane on some specified date together with the direction of the dynamical equinox on that date, neglecting nutation. Thus, the mean equator and equinox are affected only by precession. Positions in star catalogs have traditionally been referred to as a catalog equator and equinox that approximates the mean equator and equinox of a standard epoch. (See catalog equinox; true equator and equinox.)

mean motion

In undisturbed elliptic motion, the constant angular speed required for a body to complete one revolution in an orbit of a specified semimajor axis.

mean place

The coordinates of a star or other celestial object (outside the solar system) at a specific date, in the Barycentric Celestial Reference System (BCRS). Conceptually, the coordinates represent the direction of the object as it would hypothetically be observed from the solar system barycenter at the specified date, with respect to a fixed coordinate system (e.g., the axes of the International Celestial Reference Frame (ICRF)), if the masses of the Sun and other solar system bodies were negligible.

mean solar time

A measure of time based conceptually on the diurnal motion of a fiducial point, called the fictitious mean Sun, with uniform motion along the celestial equator.

meridian

A great circle passing through the celestial poles and through the zenith of any location on Earth. For planetary observations a meridian is half the great circle passing through the planet's poles and through any location on the planet.

month

The period of one complete synodic or sidereal revolution of the Moon around the Earth; also, a calendrical unit that approximates the period of revolution.

Moonrise, Moonset

The times at which the apparent upper limb of the Moon is on the astronomical horizon; i.e., when the true zenith distance, referred to the center of the Earth, of the central point of the disk is $90° 34' + s - \pi$, where s is the Moon's semidiameter, π is the horizontal parallax, and 34' is the adopted value of horizontal refraction.

nadir

The point on the celestial sphere diametrically opposite to the zenith.

node

Either of the points on the celestial sphere at which the plane of an orbit intersects a reference plane. The position of one of the nodes (the longitude of the ascending node is traditionally used as one of the standard orbital elements.)

nutation

The oscillations in the motion of the rotation pole of a freely rotating body that is undergoing torque from external gravitational forces. Nutation of the Earth's pole is specified in terms of components in obliquity and longitude. (See longitude, celestial.)

obliquity

In general, the angle between the equatorial and orbital planes of a body or, equivalently, between the rotational and orbital poles. For the Earth the obliquity of the ecliptic is the angle between the planes of the equator and the ecliptic.

occultation

The obscuration of one celestial body by another of greater apparent diameter; especially the passage of the Moon in front of a star or planet, or the disappearance of a satellite behind the disk of its primary. If the primary source of illumination of a reflecting body is cut off by the occultation, the phenomenon is also called an eclipse. The occultation of the Sun by the Moon is a solar eclipse. (See eclipse, solar.)

opposition

A configuration of the Sun, Earth and a planet in which the apparent geocentric longitude of the planet differs by 180° from the apparent geocentric longitude of the Sun. (See longitude, celestial.)

orbit

The path in space followed by a celestial body as a function of time. (See orbital elements.)

orbital elements

A set of six independent parameters that specifies an instantaneous orbit. Every real orbit can be represented as a sequence of instantaneous ellipses of hyperbolae sharing one of their foci. At each instant of time, the position and velocity of the body is characterized by its place on one such instantaneous curve. The evolution of this representation is mathematically described by evolution of the values of orbital elements. Different sets of geometric parameters may be chosen to play the role of orbital elements. The set of Keplerian elements is one of many such sets. When the Lagrange constraint (the requirement that the instantaneous orbit is tangential to the actual orbit) is imposed upon the orbital elements, they are called osculating-elements.

osculating elements

A set of parameters that specifies the instantaneous position and velocity of a celestial body in its perturbed orbit. Osculating elements describe the unperturbed (two-body) orbit that the body would follow if perturbations were to cease instantaneously. (See orbital elements; instantaneous orbit.)

parallax

The difference in apparent direction of an object as seen from two different locations; conversely, the angle at the object that is subtended by the line joining two designated points. Geocentric (diurnal) parallax is the difference in direction between a topocentric observation and a hypothetical geocentric observation. Heliocentric or annual parallax is the difference between hypothetical geocentric and heliocentric observations; it is the angle subtended at the observed object by the semimajor axis of the Earth's orbit. (See also horizontal parallax.)

parsec

The distance at which one astronomical unit (a.u.) subtends an angle of one second of arc; equivalently the distance to an object having an annual parallax of one second of arc.

penumbra

The portion of a shadow in which light from an extended source is not completely cut off by an intervening body; the area of partial shadow surrounding the umbra.

pericenter

The point in an orbit that is nearest to the center of force. (See perigee; perihelion.)

perigee

The point at which a body around the Earth is closest to the Earth. Perigee is sometimes used with reference to the apparent orbit of the Sun around the Earth.

perihelion

The point at which a body in orbit around the Sun most closely approaches the Sun.

period

The interval of time required to complete one revolution in an orbit or one cycle of a periodic phenomenon, such as a cycle of phases. (See phase.)

perturbations

Deviations between the actual orbit of a celestial body and an assumed reference orbit; also, the forces that cause deviations between the actual and reference orbits. Perturbations, according to the first meaning, are usually calculated as quantities to be added to the coordinates of the reference orbit to obtain the precise coordinates.

phase

The name applied to the apparent degree of illumination of the disk of the Moon or a planet as seen from Earth (crescent, gibbous, full, etc.). Numerically, the ratio of the illuminated area of the apparent disk of a celestial body to the entire area of the apparent disk; i.e., the fraction illuminated. Phase is also used, loosely, to refer to one aspect of an eclipse (partial phase, annular phase, etc.). (See lunar phases.)

phase angle

The angle measured at the center of an illuminated body between the light source and the observer.

photometry

A measurement of the intensity of light, usually specified for a specific wavelength range.

planetocentric coordinates

Coordinates for general use, where the z-axis is the mean axis of rotation, the x-axis is the intersection of the planetary equator (normal to the z-axis through the center of mass) and an arbitrary prime meridian, and the y-axis completes a right-hand coordinate system. Longitude of a point is measured positive to the prime meridian as defined by rotational elements. Latitude of a point is the angle between the planetary equator and a line to the center of mass. The radius is measured from the center of mass to the surface point.

planetographic coordinates

Coordinates for cartographic purposes dependent on an equipotential surface as a reference surface. Longitude of a point is measured in the direction opposite to the rotation (positive to the west for direct rotation) from the cartographic position of the prime meridian defined by a clearly observable surface feature. Latitude of a point is the angle between the planetary equator (normal to the z-axis and through the center of mass) and normal to the reference surface at the point. The height of a point is specified as the distance above a point with the same longitude and latitude on the reference surface.

polar motion

The irregularly varying motion of the Earth's pole of rotation with respect to the Earth's crust. (See Celestial Ephemeris Pole.)

position angle

The relative orientation of two celestial bodies on the sky. Position angle is measured east of north from 0 degrees through to 360 degrees. For example, if object B has a position angle of 90 degrees relative to object A then object B is directly east of object A.

precession

The uniformly progressing motion of the pole of rotation of a freely rotating body in a complex (nonprincipal) spin state. Precession is caused by a singular event (a collision or a progenitor's disruption, or a tidal interaction at a close approach) or by prolonged influence (jetting, in the case of comets, or continuous torques, in the case of planes). In the case of the Earth, the component of precession caused mainly by the Sun and Moon acting on the Earth's equatorial bulge is called lunisolar precession. The motion of the ecliptic due to the action of the planets is called planetary precession (i.e., it is a precession of the Earth's orbital plane), and the sum of lunisolar and planetary precession is called general precession. (See nutation.)

proper motion

The projection onto the celestial sphere of the space motion of a star relative to the solar system; thus, the transverse component of the space motion of a star with respect to the solar system. Proper motion is usually tabulated in star catalogs as changes in right ascension and declination per year or century.

quadrature

A configuration in which two celestial bodies have apparent longitudes that differ by 90° as viewed from a third body. Quadratures are usually tabulated with respect to the Sun as viewed from the center of the Earth. (See longitude, celestial.)

radial velocity

The rate of change of the distance to an object.

refraction, astronomical

The change in direction of travel (bending) of a light ray as it passes obliquely through the atmosphere. As a result of refraction the observed altitude of a celestial object is greater than its geometric altitude. The amount of refraction depends on the altitude of the object and on atmospheric conditions.

retrograde motion

For orbital motion in the solar system, motion that is clockwise in the orbit as seen from the north pole of the ecliptic; for an object observed on the celestial sphere, motion that is from east to west, resulting from the relative motion of the object and the Earth. (See direct motion.)

right ascension

Angular distance on the celestial sphere measured eastward along the celestial equator from the equinox to the hour circle passing through the celestial object. Right ascension is usually given in combination with declination.

second, Système International (SI)

The duration of 9 192 631 770 cycles of radiation corresponding to the transition between two hyperfine levels of the ground state of cesium 133.

selenocentric

With reference to, or pertaining to, the center of the Moon.

semidiameter

The angle at the observer subtended by the equatorial radius of the Sun, Moon or a planet.

semimajor axis

Half the length of the major axis of an ellipse; a standard element used to describe an elliptical orbit. (See elements, orbital.)

sidereal day

The interval of time between two consecutive transits of the catalog equinox. (See sidereal time.)

sidereal hour angle

Angular distance on the celestial sphere measured westward along the celestial equator from the catalog equinox to the hour circle passing through the celestial object. It is equal to 360° minus right ascension in degrees.

sidereal time

The measure of time defined by the apparent diurnal motion of the catalog equinox; hence, a measure of the rotation of the Earth with respect to the stars rather than the Sun.

solstice

Either of the two points on the ecliptic at which the apparent longitude of the Sun is 90° or 270°; also the time at which the Sun is at either point. (See longitude, celestial.)

spectral types or classes

The categorization of stars according to their spectra, primarily due to differing temperatures of the stellar atmosphere. From the hottest to the coolest, the spectral types are O, B, A, F, G, K and M.

standard epoch

A date and time that specifies the reference system to which celestial coordinates are referred. (See mean equator and equinox.)

stationary point

The time or position at which the rate of change of the apparent right ascension of a planet is momentarily zero. (See apparent place.)

sunrise, sunset

The times at which the apparent upper limb of the Sun is on the astronomical horizon; i.e., when the true zenith distance, referred to the center of the Earth, of the central point of the disk is 90°50', based on adopted values of 34' for horizontal refraction and 16' for the Sun's semidiameter.

surface brightness (of a planet)

The visual magnitude of an average square arcsecond area of the illuminated portion of the apparent disk of the Moon or a planet.

synodic period

The mean interval of time between successive conjunctions of a pair of planets, as observed from the Sun; or the mean interval between successive conjunctions of a satellite with the Sun, as observed from the satellite's primary.

synodic time

Pertaining to successive conjunctions; successive returns of a planet to the same aspect as determined by Earth.

TAI

See International Atomic Time (TAI).

TCB

See Barycentric Coordinate Time (TCB).

TCG

See Geocentric Coordinate Time (TCG).

TDB

See Barycentric Dynamical Time (TDB).

T_{eph}

The independent argument of the JPL planetary and lunar ephermerides DE405/LE405; in the terminology of General Relativity, a barycentric coordinate time scale. Teph is a linear function of the TCB and has the same rate ac TT over the time span of the ephemeris. In this volume, Teph is regarded as functionally equivalent to TDB. (See Barycentric Coordinate Time (TCB); Terrestrial Time (TT); Barycentric Dynamical Time (TDB)).

Terrestrial Time (TT)

An idealized form of International Atomic Time (TAI) with an epoch offset; in practice TT = TAI + $32^{s}184$. TT thus advances by SI seconds on the geoid. Used as an independent argument for apparent geocentric ephemerides. (See second, Système International.)

terminator

The boundary between the illuminated and dark areas of the apparent disk of the Moon, a planet or a planetary satellite.

Time Zone

The Standard Time within each time zone is offset from the Coordinated Universal Time by a constant number of hours, where Standard Time = UT + time zone offset. Standard time in the U.S. and its territories is observed within eight time zones and the Time Zone offset is an integral number of hours. The time zone offset for other non-U.S. locations may be fractional number of hours (e.g., 8h 30 min.).

topocentric

With reference to, or pertaining to, a point on the surface of the Earth.

transit

The passage of the apparent center of the disk of a celestial object across a meridian; also, the passage of one celestial body in front of another of greater apparent diameter (e.g., the passage of Mercury or Venus across the Sun or Jupiter's satellites across its disk); however, the passage of the Moon in front of the larger apparent Sun is called an annular eclipse. The passage of a body's shadow across another body is called a shadow transit; however, the passage of the Moon's shadow across the Earth is called a solar eclipse. (See eclipse, annular; solar eclipse.)

true anomaly

The angle, measured at the focus nearest the pericenter of an elliptical orbit, between the pericenter and the radius vector from the focus to the orbiting body; one of the standard orbital elements. (See orbital elements; eccentric anomaly; mean anomaly.)

true equator and equinox

The celestial coordinate system determined by the instantaneous positions of the celestial equator and ecliptic. The motion of this system is due to the progressive effect of precession and the short-term, periodic variations of nutation. (See mean equator and equinox.)

twilight

The interval of time preceding sunrise and following sunset during which the sky is partially illuminated. Civil twilight comprises the interval when the zenith distance, referred to the center of the Earth, of the central point of the Sun's disk is between 90° 50' and 96°, nautical twilight comprises the interval from 96° to 102°, astronomical twilight comprises the interval from 102° to 108°. (See sunrise, sunset.)

umbra

The portion of a shadow cone in which none of the light from an extended light source (ignoring refraction) can be observed.

Universal Time (UT)

A generic reference to one of several time scales that approximate the mean diurnal motion of the Sun; loosely, mean solar time on the Greenwich meridian (previously referred to as Greenwich Mean Time). In current usage, UT refers either to a time scale called UT1 or to Coordinated Universal Time (UTC); in this volume, UT always refers to UT1. UT1 is formally defined by a mathematical expression that relates it to sidereal time. Thus, UT1 is observationally determined by the apparent diurnal motions of celestial bodies, and is affected by irregularities in the Earth's rate of rotation. UTC is an atomic time scale but is maintained within $0\overset{s}{.}9$ of UT1 by the introduction (or removal) of 1-second steps when necessary. (See leap second.)

UTC

See Coordinated Universal Time (UTC).

vernal equinox

The ascending node of the ecliptic on the celestial equator; also the time at which the apparent longitude of the Sun is 0°. (See apparent place; longitude, celestial; equinox.)

vertical

Apparent direction of gravity at the point of observation (normal to the plane of a free level surface.)

week

An arbitrary period of days, usually seven days; approximately equal to the number of days counted between the four phases of the Moon. (See lunar phases.)

year

A period of time based on the revolution of the Earth around the Sun. The calendar year is an approximation to the tropical year. The anomalistic year is the mean interval between successive passages of the Earth through perihelion. The sidereal year is the mean period of revolution with respect to the background stars. (See Gregorian calendar; year, tropical; Julian year.)

year, Besselian

The period of one complete revolution in right ascension of the fictitious mean Sun, as defined by Newcomb.

year, Julian

See Julian year.

year, tropical

The period of one complete revolution of the mean longitude of the Sun with respect to the dynamical equinox. The tropical year comprises a complete cycle of seasons, and its length is approximated in the long term by the civil (Gregorian) calendar.

zenith

In general, the point directly overhead on the celestial sphere. The astronomical zenith is the extension to infinity of a plumb line. The geocentric zenith is defined by the line from the center of the Earth through the observer. The geodetic zenith is the normal to the geodetic ellipsoid at the observer's location.

zenith distance

Angular distance on the celestial sphere measured along the great circle from the zenith to the celestial object. Zenith distance is 90° minus altitude.

Index